Applied Speech Processing

PRIMERS IN BIOMEDICAL IMAGING
DEVICES AND SYSTEMS SERIES

Applied Speech Processing

Algorithms and Case Studies

Edited by

NILANJAN DEY

Department of Computer Science and Engineering,
JIS University, Kolkata, India

ELSEVIER

ACADEMIC PRESS

An imprint of Elsevier

Academic Press
125 London Wall, London EC2Y 5AS, United Kingdom
525 B Street, Suite 1650, San Diego, CA 92101, United States
50 Hampshire Street, 5th Floor, Cambridge, MA 02139, United States
The Boulevard, Langford Lane, Kidlington, Oxford OX5 1GB, United Kingdom

© 2021 Elsevier Inc. All rights reserved.

No part of this publication may be reproduced or transmitted in any form or by any means, electronic
or mechanical, including photocopying, recording, or any information storage and retrieval system,
without permission in writing from the publisher. Details on how to seek permission, further
information about the Publisher's permissions policies and our arrangements with organizations
such as the Copyright Clearance Center and the Copyright Licensing Agency, can be found at our
website: www.elsevier.com/permissions.

This book and the individual contributions contained in it are protected under copyright by the
Publisher (other than as may be noted herein).

Notices
Knowledge and best practice in this field are constantly changing. As new research and experience
broaden our understanding, changes in research methods, professional practices, or medical
treatment may become necessary.

Practitioners and researchers must always rely on their own experience and knowledge in
evaluating and using any information, methods, compounds, or experiments described herein.
In using such information or methods they should be mindful of their own safety and the safety
of others, including parties for whom they have a professional responsibility.

To the fullest extent of the law, neither the Publisher nor the authors, contributors, or editors, assume
any liability for any injury and/or damage to persons or property as a matter of products liability,
negligence or otherwise, or from any use or operation of any methods, products, instructions, or
ideas contained in the material herein.

Library of Congress Cataloging-in-Publication Data
A catalog record for this book is available from the Library of Congress

British Library Cataloguing-in-Publication Data
A catalogue record for this book is available from the British Library

ISBN: 978-0-12-823898-1

For information on all Academic Press publications
visit our website at https://www.elsevier.com/books-and-journals

Publisher: Mara Conner
Acquisitions Editor: Fiona Geraghty
Editorial Project Manager: Chris Hockaday
Production Project Manager: Prem Kumar Kaliamoorthi
Cover Designer: Mark Rogers

Typeset by SPi Global, India

Working together
to grow libraries in
developing countries

www.elsevier.com • www.bookaid.org

Contents

Contributors

Mohammed A.M. Abdullah
Computer and Information Engineering Department, College of Electronics Engineering, Ninevah University, Mosul, Iraq

Musab T.S. Al-Kaltakchi
Department of Electrical Engineering, College of Engineering, Mustansiriyah University, Baghdad, Iraq

Faramarz Alsharif
Kitami Institute of Technology, Kitami, Hokkaido, Japan

Mohammad Reza Alsharif
University of the Ryukyus, Ryukyus, Okinawa, Japan

K.T. Bibish Kumar
Computer Speech & Intelligence Research Centre, Department of Physics, Government College, Madappally, Calicut, Kerala, India

A. Chandrasekar
Department of Computer Science and Engineering, St. Joseph's College of Engineering, Chennai, India

Satnam S. Dlay
School of Electrical and Electronic Engineering, Newcastle University, Newcastle upon Tyne, United Kingdom

R. Hariprasad
Zoho Corporation, Chennai, India

Sunil John
Computer Speech & Intelligence Research Centre, Department of Physics, Government College, Madappally, Calicut, Kerala, India

M. Kalamani
Department of Electronics and Communication Engineering, Velalar College of Engineering and Technology, Erode, India

Muhammad Irfan Khattak
Department of Electrical Engineering, University of Engineering & Technology, Peshawar, Pakistan

Mahdi Khosravy
Media Integrated Communication Laboratory, Graduate School of Engineering, Osaka University, Suita, Osaka, Japan

M. Krishnamoorthi
Department of Computer Science and Engineering, Dr. N.G.P. Institute of Technology, Coimbatore, India

K.M. Muraleedharan
Computer Speech & Intelligence Research Centre, Department of Physics, Government College, Madappally, Calicut, Kerala, India

S. Radhika
Department of Electrical and Electronics Engineering, School of Electrical and Electronics Engineering, Sathyabama Institute of Science and Technology, Chennai, India

Hamurabi Gamboa Rosales
Department of Signal Processing and Acoustics, Faculty of Electrical Engineering, Autonomous University of Zacatecas, Zacatecas, Mexico

Nasir Saleem
Department of Electrical Engineering, Faculty of Engineering & Technology, Gomal University, Dera Ismail Khan; Department of Electrical Engineering, University of Engineering & Technology, Peshawar, Pakistan

J. Sangeetha
Srinivasa Ramanujan Centre, SASTRA Deemed University, Kumbakonam, India

S. Subhiksha
Department of IT, School of Computing, SASTRA Deemed University, Tirumalaisamudhram, India

R.K. Sunil Kumar
School of Information Science and Technology, Kannur University, Kannur, Kerala, India

Navneet Upadhyay
Department of Electronics and Communication Engineering, The LNM Institute of Information Technology, Jaipur, India

Wai L. Woo
Department of Computer and Information Sciences, Northumberia University, Newcastle upon Tyne, United Kingdom

Linnan Zhang
NTT Data Global Service, Tokyo, Japan

Preface

This book presents basics of speech data analysis and management tools with several applications by covering different techniques in speech signal processing. Part 1 "Speech enhancement and synthesis" includes five chapters, and Part 2 "Speech identification, feature selection, and classification" includes three chapters. In Chapter 1, Radhika and Chandrasekar apply a data-selective affine projection algorithm (APA) for speech processing applications. To remove noninnovative data and impulsive noise, the authors propose a kurtosis of error-based update rule. Results of the author's study show that the proposed scheme is suitable for speech processing application, as it obtained reduction in space and time as well as increased efficiency. In Chapter 2, Upadhyay and Rosales propose a recursive noise estimation-based Wiener filtering method for monaural speech enhancement. This method estimates the noise from present and past frames of noisy speech continuously, using a smoothing parameter value between 0 and 1. The authors compare the performance of the proposed approach with traditional speech enhancement methods. In Chapter 3, Kalamani and Krishnamoorthi develop a least mean square adaptive noise reduction (LMS-ANR) algorithm for enhancing the Tamil speech signal with acceptable quality under a nonstationary noisy environment that automatically adapts its coefficients with respect to input noisy signals. In Chapter 4, Saleem and Khattak propose an unsupervised speech enhancement to decrease the noise in nonstationary and difficult noisy backgrounds. They accomplish this by replacing the spectral phase of the noisy speech with an estimated spectral phase and merging it with a novel time-frequency mask during signal reconstruction. The results show considerable improvements in terms of short-time objective intelligibility (STOI), perceptual evaluation of speech quality (PESQ), segmental signal-to-noise ratio (SSNR), and speech distortion. Chapter 5 by Khosravy et al. introduces a novel approach to speech synthesis by adaptively constructing and combining the harmonic components based on the fusion of Fourier series and adaptive filtering.

Part 2 begins with Chapter 6 by Bibish Kumar et al., who discuss the primary task of identifying visemes and the number of frames required to encode the temporal evolution of vowel and consonant phonemes using an audio-visual Malayalam speech database. In Chapter 7, Al-Kaltakchi et al. propose novel fusion strategies for text-independent speaker identification. The authors apply four main simulations for speaker identification accuracy (SIA), using different fusion strategies, including feature-based early fusion, score-based late fusion, early-late fusion (combination of feature and score-based), late fusion for concatenated features, and statistically independent normalized scores fusion for all the previous scores. In Chapter 8, Sangeetha et al. use the TAU Urban Acoustic Scenes 2019 dataset and DCASE 2016 Challenge Dataset to compare various standard classifications, including support vector machines (SVMs) using different kernels, decision trees, and logistic regression for classifying audio events. The authors extract several features to generate the feature vector, such as Mel-frequency cepstral coefficients (MFCCs). The experimental results prove that the SVM with linear kernels yields the best result compared to other machine learning algorithms.

The editor would like to express his gratitude to the authors and referees for their contributions. Without their hard work and cooperation, this book would not have come to fruition. Extended thanks are given to the members of the Elsevier team for their support.

Nilanjan Dey, Editor
Department of Computer Science and Engineering,
JIS University, Kolkata, India

PART 1

Speech enhancement and synthesis

CHAPTER 1

Kurtosis-based, data-selective affine projection adaptive filtering algorithm for speech processing application

S. Radhika[a] and A. Chandrasekar[b]
[a]Department of Electrical and Electronics Engineering, School of Electrical and Electronics Engineering, Sathyabama Institute of Science and Technology, Chennai, India
[b]Department of Computer Science and Engineering, St. Joseph's College of Engineering, Chennai, India

1.1 Introduction

Speech is an important mode of communication that is produced naturally without any electronic devices. Some of the major applications of the speech signal include vehicle automation, gaming, communication systems, new language acquisition, correct pronunciation, online teaching, and so on. In addition, speech signals also find applications in medicine for developing assistive devices, identifying cognitive disorders, and so on. In the military field, speech signals are used for the development of high-end fighter jets, immersive audio flights, and more [1–4]. Nowadays, speech signal processing is inevitable for the development of smart cities. Generally, these speech signals are collected using acoustic sensors deployed in different places. The tremendous increase in sensors at cheaper rates results in the availability of large amounts of data. The bulk data sets produced by these sensors demand huge amounts of memory space and fast processing speeds. Moreover, these aggregated data sets often suffer from outliers that may occur due to the surroundings or measurement errors [5]. Another key issue is that not all data in the data set are useful, as some data may not contain information. Thus there is a growing demand for some sort of adaptive algorithm that incorporates data selectivity with the capability to remove noise and outliers [2]. Basically, error is used as a metric to conclude the level of new information in the data set. As the speech signal contains more non-Gaussian noise, second-order statistics of error are not suitable metrics for speech processing applications.

Applied Speech Processing
https://doi.org/10.1016/B978-0-12-823898-1.00005-9

© 2021 Elsevier Inc.
All rights reserved.

This work proposes an improved data-selective affine projection algorithm (APA) based on kurtosis of error for speech processing applications.

This chapter is organized as follows. Section 1.2 discusses the nature of speech signals, adaptive algorithms for speech processing applications, the traditional adaptive algorithms of Least Mean Square (LMS) and Normalized LMS (NLMS), and the proposed APA algorithm. It also examines the problems associated with current data-selective adaptive algorithms. Section 1.3 details the system model for the adaptive algorithm, and Section 1.4 examines the proposed update rule. Section 1.4 also discusses the mean squared error (MSE) of the algorithm and the nature of noise and error sources. Further, the section analyzes the steady-state MSE of the APA algorithm using the new proposed update rule. It provides simulations in which different scenarios are taken and compared with their original counterparts, and discusses the results obtained. Finally, Section 1.5 presents conclusions along with the limitations and future scope of the proposed work.

1.2 Literature review

In order to design adaptive algorithms suitable for speech processing applications, it is required to understand their nature. The impulse response of a speech signal has the general characteristics of long sequence length. It is also said to be time varying and subjected to both impulsive and background noises. Thus it is evident that a filter capable of adjusting the filter coefficients according to the change of signal properties is required for speech processing applications, as the signal statistics are not known prior or are time varying. An adaptive filter is a type of filter in which the coefficients are changed depending on the adaptive algorithm used. Therefore adaptive filters are the unanimous choice for speech processing applications [6]. The criteria to be satisfied by an adaptive filter used for speech are data-selective capability, fast convergence, low steady-state error, robustness against background and impulsive noise (in case of double talk), and reduced computational complexity [7].

Traditionally, LMS and NLMS adaptive filters and their variants have been widely used because of their simplicity and easy implementation [8]. The main function of these algorithms is to estimate speech in such a way that MSE is minimized. Since the speech signal is said to be time varying and requires a long-length adaptive filter with hundreds of filter coefficients and a constant step-size parameter to update the filter coefficients, there is performance trade-off in terms of convergence speed and steady-state

MSE. As such, LMS and NLMS perform poorly for speech signals [9]. APA is better for speech processing applications as it provides faster convergence with lower steady-state MSE at the cost of increased computational complexity. The conventional adaptive algorithms do not exhibit data selectivity and hence they are not suitable for speech processing applications. In order to obtain the best performance, several variants were developed. Set membership adaptive algorithms were found be an important family of data-selective adaptive filters [10–13]. The probability-based update rule is not suitable when the outliers are not Gaussian, as they are modeled as Gaussian distribution. These algorithms also lack in the number of solutions involved because they update the coefficients in blocks and not as a single scalar value. Later, data-censoring adaptive algorithms were developed [14,15]. The idea behind these algorithms is that noninnovative and noise data are censored to participate in the update recursion, which is found to be useful in wireless and ad hoc networks, medical data sets, and large data applications. In data-selective adaptive algorithms, thresholds are selected to propose the probability of update [16–19]. The threshold values are based on modeling using Gaussian distribution, which is more suitable if the error is said to be Gaussian. However, in many real-time applications, the error is found to have heavy- and light-tailed distributions [20].

Statistical evidence indicates that higher-order statistics measure the error more accurately than second-order statistics. The literature provides evidence that kurtosis of error performs better than MSE when the error is non-Gaussian, especially when the error tends to have more heavy- and light-tailed noise sources [14,15,19]. Thus the existing rule is not suitable when the outliers are not Gaussian. Therefore there is no single strategy available to tackle large volume, noise-prone data. This chapter fills the research gap by designing and developing a data–selective adaptive algorithm that selects the thresholds suitable for speech processing applications. This is the motivation behind the proposed update rule in this work. Here the two thresholds are selected based on second order and kurtosis of error, respectively. Thus the proposed data-selective APA can obtain the same performance as that of DS-APA even when they are subjected to non-Gaussian noise with lesser data sets.

1.3 System model

In order to formulate the problem, let us consider Fig. 1.1 in which the system is assumed to be an unknown system and the adaptive filter is used for an

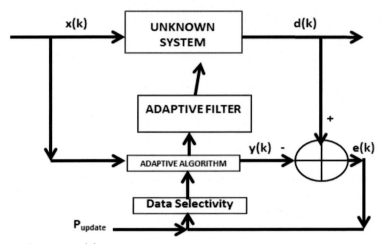

Fig. 1.1 System model.

unknown system identification problem. The optimal filter weights w_o are given by

$$w_o = [w_o, w_1, w_2, ..., w_{N-1}]^T \tag{1.1}$$

Let $x(k)$ be the input to the system given by

$$x(k) = [x(k), x(k-1), x(k-2), ..., x(k-N+1)]^T \tag{1.2}$$

where k is time index and N is the length of the input and the weights. The desired response is given by $d(k)$, which is modeled as a linear regression model given by

$$d(k) = w_o^T x(k) + n(k) \tag{1.3}$$

where $n(k)$ is the noise signal, which is made of both background noise $v(k)$ with variance σ_v^2 and impulsive noise $u(k)$ with variance σ_u^2. The desired response in vector form is given by

$$d(k) = [d(k)\, d(k-1)...d(k-P+1)]^T \tag{1.4}$$

The adaptive filter is used to estimate the weights for the given input $x(k)$. Let $w(k)$ be the estimated weight vector given by

$$w(k) = [w(k), w(k-1), ..., w(k-N+1)]^T \tag{1.5}$$

The estimated response is given by

$$y(k) = \mathbf{A}(k)\mathbf{w}(k) \tag{1.6}$$

where $\mathbf{A}(k) = [x(k)\ x(k-1)\ x(k-2)...x(k-P+1)]^T$ is the projection vector obtained by taking the delayed version of input vector and P is projection order. Usually P is less than or equal to N. The error signal $e(k)$ is given by $e(k) = d(k) - y(k)$. The vector form of error is given by

$$e(k) = [e(k), e(k-1), ..., e(k-P+1)]^T \tag{1.7}$$

The error vector is used for updating the filter coefficients in the adaptive filter using the adaptive algorithm. The update recursion of conventional APA [3] is given by

$$w(k+1) = w(k) + \mu A^T(k)(\delta I + A(k)A^T(k))^{-1} e(k) \tag{1.8}$$

Here I is the identity matrix of order $P \times P$, μ is the step size.

The following are the assumptions used for APA.

A1: The noise sequence $n(k)$ is assumed to be identically and independently distributed (i.d.d.) and is statistically independent of the regression data $A(k)$.

A2: The dependency of weight error vector $\tilde{w}(k)$ on the past noise is ignored.

A3: The noise sequence and the error are uncorrelated.

1.4 Proposed update rule

This section deals with the proposed update rule. If the error value is within the acceptable range, then it indicates that the coefficients do not bring new improvement to the algorithm. Thus it is not required to update the coefficients, as they do not take part in bringing new data. If the error is more than the upper acceptable limit, those data are considered as outliers and they need to be removed. This strategy is now applied for APA. P_{update} is the probability of update, which ranges from 0 to 1. It illustrates the amount of update in the algorithm in order to make the estimate of the filter coefficients.

If the weight error vector is given as $\tilde{w}(k) = w_0 - w(k)$, then the MSE is written as $E[\|e(k)\|^2]$. If the error signal and the noise are uncorrelated, we obtain

$$E[\|e(k)\|^2] = \sigma_n^2 + \varepsilon_{excess}(k) \tag{1.9}$$

where $\varepsilon_{excess}(k)$ is the excess mean square error and σ_n^2 is the variance of the noise source, respectively. Eq. (1.9) is very important for the adaptive algorithm because it is used for updating filter coefficients.

Consider Fig. 1.1 again in which the estimated output is obtained at a time instant k. Let the error measured be $e(k)$. Let the MSE be obtained at the time instant k. If the MSE is smaller than some value times of variance of noise (i.e., $\|e(k)\|^2 \leq \gamma_{\min}\sigma_n^2$), then it implies that there is no new information in the data set. Thus the updating need not be made or $w(k+1) = w(k)$. On the other hand, if the mean square value of error is greater than maximum value times of noise $\|e(k)\|^2 > \gamma_{\max}\sigma_n^2$ it implies that an outlier such as measurement noise would have occurred and thus the update is not done and the data can be discarded [16].

If the error signal is substituted in Eq. (1.9) and if $E\|e(k)\|^2 = \sigma_e^2$ is applied, using the orthogonality principle the MSE can be written as $\sigma_n^2 + \lambda_{\min}E\|\tilde{w}(\infty)\|^2 \leq \sigma_e^2 \leq \sigma_n^2 + \lambda_{\max}E\|\tilde{w}(\infty)\|^2$ where $\lambda_{\min}$ and $\lambda_{\max}$ are the minimum and maximum Eigen values. Thus under steady-state condition, Eq. (1.9) can be written as $\|e(k)\|^2 = \sigma_n^2 + \beta\sigma_n^2 = \gamma\sigma_n^2$ where $\beta\sigma_n^2$ is $\varepsilon_{excess}(\infty)$. Thus the probability-based update rule can be formulated as follows:

$$P_{update}(k) = P\left((\gamma_{\min}\sigma_n^2 \leq \|e(k)\|^2 \leq \gamma_{\max}\sigma_n^2\right) \qquad (1.10)$$

Here the l_2 norm strategy is adopted for choosing the update rule. The probability of update is said to be time varying as it depends on the value of the error. Using probability theory [21], Eq. (1.10) can be rewritten as

$$P_{update}(k) = P\left(\|e(k)\|^2 > (\gamma_{\min}\sigma_n^2)\right) - P\left(\|e(k)\|^2 > (\gamma_{\max}\sigma_n^2)\right) \qquad (1.11)$$

As the noise consists of heavy- and light-tailed noise sources, as seen in Fig. 1.2, mere Gaussian distribution cannot be used for modeling the probability. As the update rule is based on $\|e(k)\|^2$, chi square distribution, which is a special case of Gamma distribution, is used to model probability of update. Using the complementary cumulative distribution function [22], the update probability is written as

$$P_{update}(k) = \frac{\daleth_i\left(\dfrac{\gamma_{\min}\sigma_n^2|S|}{2\sigma_e^2}\right)}{\daleth\left(\dfrac{|S|}{2}\right)} - \frac{\daleth_i\left(\dfrac{\gamma_{\max}\sigma_n^2|S|}{2\sigma_e^2}\right)}{\daleth\left(\dfrac{|S|}{2}\right)} \qquad (1.12)$$

where $\daleth$ is the standard Gamma function, $\daleth_i$ is the incomplete gamma function given by $\daleth_i(x) = \int_x^\infty t^{\frac{|S|}{2}-1}e^{-t}dt$, and S is the degree of freedom.

Thus if the outlier is found to be small, then the minimum value of threshold [20] can be found as

$$\gamma_{\min} = \frac{2(1+\beta)}{|S|}\daleth_i^{-1}\left(P_{update}\daleth\left(\frac{|S|}{2}\right)\right) \qquad (1.13)$$

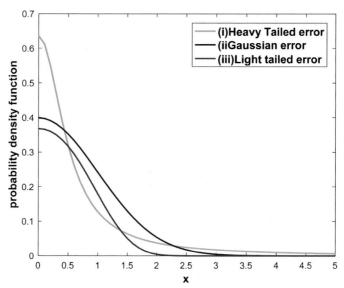

Fig. 1.2 Probability distribution of heavy-tailed, light-tailed, and Gaussian error.

As it is well known that any non–Gaussian source can be approximated as a linear combination of Gaussian distribution and chi distributions [23], the Gaussian condition can be set as a testing condition for selection of upper threshold γ_{max}. Thus if the error obtained follows Gaussian distribution, then the update rule based on MSE of error provides good performance. As seen in Fig. 1.2, if the density of error deviates from the shape of Gaussian distribution, mere MSE does not provide a feasible solution. If the error has heavy tails, then the γ_{max} should be chosen lesser than MSE so that the non-relevant information resulting from measurement error or outliers can be removed. The converse is true for light-tailed noise. Therefore an additional parameter is required to adjust γ_{max} along with the MSE. From Refs. [23–25] it is found that using higher-order statistics of error can depict the distribution with heavy tails for heavy-tailed noise and light tails for light–tailed noise with respect to normal distribution. Kurtosis is the measure of the fourth power of coefficient, which is given by [23] as

$$\text{Kurtosis} = k_e = \frac{E[e^4(n)]}{E^2[e^2(n)]} \tag{1.14}$$

Kurtosis has a greater value for sharp and heavy tails and a lesser value for light tails. Thus when used along with kurtosis of Gaussian distribution with value equal to 3, the maximum value of threshold can be effectively fixed. The value of upper threshold γ_{max} is very important as it decides the elimination of the outlier. Therefore the kurtosis of error as a number or ratio with respect to kurtosis of Gaussian error is used for setting up the upper threshold $\gamma_{\max}$. Hence the ratio $\frac{k_e}{k_G}$, where k_e is the kurtosis of error and k_G is the kurtosis of Gaussian distribution, is used along with MSE. Thus it can be seen that whenever the error contains heavy-tailed noise, then the k_e is a large value. The ratio is larger as $k_G = 3$. Hence the range of the error updating is reduced and the heavy-tailed outlier is prevented in the filter coefficients. On the other hand, if the error has light-tailed noise, the range of updating should be large so as to prevent loss of useful information. As expected for light-tailed error, the value of k_e is small and hence the ratio is smaller. Thus the proposed rule selectively operates on the data and removes all types of outliers. From Eq. (1.15), in case of outliers, the probability of update is reduced by Eq. (1.15), as they are discarded.

$$\gamma_{\max} = \frac{k_e}{k_G}\frac{2(1+\beta)}{|S|}1_i^{-1}\left(P_{update}1\left(\frac{|S|}{2}\right)\right) \tag{1.15}$$

As the error is modeled as chi-square distribution, the kurtosis of error is found to be $\frac{12}{|S|}$. Thus Eq. (1.15) can be further written as $\gamma_{\max} = \frac{12}{3|S|}\frac{2(1+\beta)}{|S|}1_i^{-1}\left(P_{update}1\left(\frac{|S|}{2}\right)\right)$.

1.4.1 Steady-state mean square analysis

The update recursion of the proposed DS-APA is written as follows:

$$w(k+1)=\begin{cases} w(k)+\mu A^T(k)\left(\delta I+A(k)A^T(k)\right)^{-1}e(k) & \text{if } \gamma_{\min}\sigma_n^2 \le \|e(k)\|^2 \le \gamma_{\max}\sigma_n^2 \\ w(k) & \text{otherwise} \end{cases} \tag{1.16}$$

In order to obtain steady-state MSE, Eq. (1.7) is taken. Substituting for $d(k)$ and $y(k)$ in $e(k)$ we get $e(k) = w_o^T x(k) + n(k) - w^T x(k)$. The error consists of two terms, namely, a priori error and a posteriori error vectors as follows:

$$e_a(k) = A(k)\tilde{w}(k) \tag{1.17}$$

$$e_p(k) = A(k)\tilde{w}(k+1) \tag{1.18}$$

where $\tilde{w}(k) = w_o^T - w^T(k)$ is the weights error vector. In order to resolve the nonlinearity present in Eq. (1.16), the model introduced in Section 1.2 is used in Eq. (1.16) as

$$w(n+1)=w(k)+\mu P_{update}A^T(k)\left(\delta I+A(k)A^T(k)\right)^{-1}e(k) \tag{1.19}$$

Eq. (1.19) is still more nonlinear due to the presence of P_{update}, which lies between 0 and 1 and is valid only for a certain range given by Eq. (1.10).

1.4.2 Mean analysis

In order to analyze in the mean sense, let $m(k)$ be a random variable that is used when the update condition $\gamma_{min}\sigma_n^2 \leq \|e(k)\|^2 \leq \gamma_{max}\sigma_n^2$ becomes valid. Thus

$$m(k) = \mu A^T(k)\left(\delta I + A(k)A^T(k)\right)^{-1}e(k) \tag{1.20}$$

Substituting Eq. (1.20) in Eq. (1.16) we get

$$w(k+1) = \begin{cases} w(k) + m(k) & \text{if } \gamma_{min}\sigma_n^2 \leq \|e(k)\|^2 \leq \gamma_{max}\sigma_n^2 \\ w(k) & \text{otherwise} \end{cases} \tag{1.21}$$

Let $\overline{m}(k)$ be a random variable written as

$$\overline{m}(k) = \begin{cases} m(k) & \text{if } \gamma_{min}\sigma_n^2 \leq \|e(k)\|^2 \leq \gamma_{max}\sigma_n^2 \\ 0 & \text{otherwise} \end{cases} \tag{1.22}$$

Thus the update recursion for the proposed DS–APA is given by

$$w(k+1) = w(k) + \overline{m}(k) \tag{1.23}$$

From Ref. [21] using conditional distribution, $E[X] = E[X/Y]$ $E[Y]$, $E[\overline{m}(k)]$ can be written as

$$\begin{aligned} E[\overline{m}(k)] = E[\overline{m}(k)|\{\|e(k)\|^2 \leq \gamma_{min}\sigma_v^2\}\left(1 - P_{update}(k)\right) + E[\overline{m}(n)|\|e(k)\|^2 \\ > \gamma_{min}\sigma_v^2]\left(P_{update}(k)\right) + E[\overline{m}(k)|\{\|e(k)\|^2 > \gamma_{max}\sigma_v^2\}\left(1 - P_{update}(k)\right) \\ + E[\overline{m}(k)|[\|e(k)\|^2 \leq \gamma_{max}\sigma_v^2]\left(P_{update}(k)\right) \end{aligned} \tag{1.24}$$

$$= 0 + E[m(k)|[\|e(k)\|^2 > \gamma_{min}\sigma_v^2]\left(P_{update}(k)\right) - E[m(k)|[\|e(k)\|^2 > \gamma_{max}\sigma_v^2]\left(P_{update}(k)\right) \tag{1.25}$$

$$E[\overline{m}(k)] = E[m(k)]P_{update}(k) \tag{1.26}$$

If the random variable $m(k)$ and $P_{update}(k)$ are independent, the expected value of update recursion becomes

$$E[w(k+1)] = E[w(k)] + E[\overline{m}(k)] \tag{1.27}$$

Thus it can be seen that the mean values of Eqs. (1.8), (1.27) are found to be the same. Thus Eq. (1.16) is approximately equal to Eq. (1.19) and it can be taken for further analysis.

In order to obtain the steady-state MSE, Eq. (1.19) is written in terms of weight error. Thus we obtain

$$\widetilde{\boldsymbol{w}}\,(k+1) = \widetilde{\boldsymbol{w}}\,(k) - \mu P_{update}\boldsymbol{A}^T(k)\big(\delta I + \boldsymbol{A}(k)\boldsymbol{A}^T(k)\big)^{-1}\boldsymbol{e}(k) \qquad (1.28)$$

If Eq. (1.28) is multiplied by $\boldsymbol{A}(k)$, we get

$$\boldsymbol{A}(k)\,\widetilde{\boldsymbol{w}}\,(k+1) = \boldsymbol{A}(k)$$
$$\widetilde{\boldsymbol{w}}\,(k) - \mu P_{update}\boldsymbol{A}(k)\boldsymbol{A}^T(k)\big(\delta I + \boldsymbol{A}(k)\boldsymbol{A}^T(k)\big)^{-1}\boldsymbol{e}(k) \qquad (1.29)$$

Writing Eq. (1.29) in terms of a priori and a posteriori error, we get

$$\boldsymbol{e}_p(k) = \boldsymbol{e}_a(k) - \mu P_{update}\boldsymbol{A}(k)\boldsymbol{A}^T(k)\big(\delta I + \boldsymbol{A}(k)\boldsymbol{A}^T(k)\big)^{-1}\boldsymbol{e}(k) \qquad (1.30)$$

Eliminating $(\delta I + \boldsymbol{A}(k)\boldsymbol{A}^T(k))^{-1}\boldsymbol{e}(k)$ from Eq. (1.30) as

$$\big(\delta I + \boldsymbol{A}(k)\boldsymbol{A}^T(k)\big)^{-1}\boldsymbol{e}(k) = \frac{\big(\boldsymbol{A}(k)\boldsymbol{A}^T(k)\big)^{-1}\big(\boldsymbol{e}_p(k) - \boldsymbol{e}_a(k)\big)}{\mu P_{update}} \qquad (1.31)$$

Substituting Eq. (1.31) in Eq. (1.28), we get

$$\widetilde{\boldsymbol{w}}\,(k+1) + \boldsymbol{A}^T(k)\big(\boldsymbol{A}(k)\boldsymbol{A}^T(k)\big)^{-1}\boldsymbol{e}_a(k) =$$
$$\widetilde{\boldsymbol{w}}\,(k) + \boldsymbol{A}^T(k)\big(\boldsymbol{A}(k)\boldsymbol{A}^T(k)\big)^{-1}\boldsymbol{e}_p(k) \qquad (1.32)$$

If the transpose of Eq. (1.32) is taken and multiplied with Eq. (1.32), we get

$$\big\|\widetilde{\boldsymbol{w}}(k+1)\big\|^2 + \boldsymbol{e}_a^T(k)\big(\boldsymbol{A}(k)\boldsymbol{A}^T(k)\big)^{-1}\boldsymbol{e}_a(k)$$
$$= \big\|\widetilde{\boldsymbol{w}}(k)\big\|^2 + \boldsymbol{e}_p^T(k)\big(\boldsymbol{A}(k)\boldsymbol{A}^T(k)\big)^{-1}\boldsymbol{e}_p(k) \qquad (1.33)$$

Eq. (1.33) is the energy conservation relation for the proposed DS-APA. The mean square analysis is obtained by taking expectation of Eq. (1.33) on both sides. Thus we get

$$E\Big[\big\|\widetilde{\boldsymbol{w}}(k+1)\big\|^2\Big] + E\Big[\boldsymbol{e}_a^T(k)\big(\boldsymbol{A}(k)\boldsymbol{A}^T(k)\big)^{-1}\boldsymbol{e}_a(k)\Big] =$$
$$E\Big[\big\|\widetilde{\boldsymbol{w}}(k)\big\|^2\Big] + E\Big[\boldsymbol{e}_p^T(k)\big(\boldsymbol{A}(k)\boldsymbol{A}^T(k)\big)^{-1}\boldsymbol{e}_p(k)\Big] \qquad (1.34)$$

As $k \to \infty$ when the algorithm reaches steady state, we obtain

$$E\Big[\big\|\widetilde{\boldsymbol{w}}(k+1)\big\|^2\Big] = E\Big[\big\|\widetilde{\boldsymbol{w}}(k)\big\|^2\Big] \qquad (1.35)$$

Substituting in Eq. (1.34) we get

$$E\Big[\boldsymbol{e}_a^T(k)\big(\boldsymbol{A}(k)\boldsymbol{A}^T(k)\big)^{-1}\boldsymbol{e}_a(k)\Big] = E\Big[\boldsymbol{e}_p^T(k)\big(\boldsymbol{A}(k)\boldsymbol{A}^T(k)\big)^{-1}\boldsymbol{e}_p(k)\Big] \qquad (1.36)$$

Substituting $e_p(k)$ from Eq. (1.30) in Eq. (1.36), we get

$$E\left[e_p^T(k)\left(A(k)A^T(k)\right)^{-1}e_p(k)\right] =$$

$$E\left[e_a^T(k)\left(A(k)A^T(k)\right)^{-1}e_a(k)\right] - \mu P_{update}E\left[e_a^T(k)B(k)e(k)\right] -$$

$$\mu P_{update}E\left[e^T(k)B(k)e_a(k)\right] + \mu^2 P_{update}^2 E\left[e^T(k)C(k)e(k)\right] \qquad (1.37)$$

where

$$B(k) = \left(\delta I + A(k)A^T(k)\right)^{-1} \quad \text{and}$$

$$C(k) = \left(\delta I + A(k)A^T(k)\right)^{-1}\left(A(k)A^T(k)\right)^{-1}\left(\delta I + A(k)A^T(k)\right)^{-1} \quad (1.38)$$

Substituting Eq. (1.37) in Eq. (1.36), we get

$$E\left[e_a^T(k)B(k)e(k)\right] + E\left[e^T(k)B(k)e_a(k)\right] = \mu P_{update}E\left[e^T(k)C(k)e(k)\right]$$

$$(1.39)$$

Substituting $e(k) = e_a(k) + n(k)$ in Eq. (1.39), and if assumption 1 is used, we get

$$2E\left[e_a^T(k)B(k)e_a(k)\right] = \mu P_{update}E\left[e_a^T(k)C(k)e_a(k)\right] + \mu P_{update}E\left[n^T(k)C(k)n(k)\right]$$

$$(1.40)$$

If the assumption that $A(k)$ is statistically independent of $e_a(k)$ is used along with $S = (1.1^T)$ [5], then

$$2E\left[\|e_a(k)\|^2\right]Trace(E(S \cdot B(k))) = \mu P_{update}E\left[\|e_a(k)\|^2\right]Trace(E(S \cdot C(k)))$$
$$+ \mu P_{update}\sigma_n^2 Trace(E(C(k)))$$

$$(1.41)$$

Under steady-state condition as $k \rightarrow \infty$,

$$\varepsilon_{excess}(\infty) = \beta = E\left[\|e_a(\infty)\|^2\right]$$
$$= \frac{\mu P_{update}\sigma_n^2 Trace(E(C(k)))}{2 Trace(E(S \cdot B(k))) - \mu P_{update}Trace(E(S \cdot C(k)))} \qquad (1.42)$$

Eq. (1.41) gives the steady-state excess MSE of the proposed DS-APA. If the regularization factor is small such that $Trace(E(B(k))) = Trace(E(C(k)))$, then

$$\varepsilon_{excess}(\infty) = \frac{\mu P_{update}\sigma_n^2}{(2 - \mu P_{update})}\frac{Trace(C(k))}{Trace(S \cdot C(k))} = \frac{\mu P_{update}\sigma_n^2}{(2 - \mu P_{update})}\frac{Trace(E(C(k)))}{Trace(E(C(1,1)))}$$

$$(1.43)$$

Further if
$$Trace(E(C(k))) = E\left[\frac{P}{\|x(k)^2\|}\right] \text{ and } Trace(E(C(1,1))) = Trace\ (R_{xx})$$

$$\varepsilon_{excess}(\infty) = \frac{\mu P_{update}\sigma_n^2}{(2-\mu P_{update})}\frac{Trace(C(k))}{Trace(S\cdot C(k))} = \frac{\mu P_{update}\sigma_n^2}{(2-\mu P_{update})}\frac{Trace(E(C(k)))}{Trace(E(C(1,1)))}$$

$$(1.44)$$

Substituting Eq. (1.44) in MSE, we obtain the steady-state MSE as

$$E\left[\|e(k)\|^2\right] = \sigma_n^2 + \varepsilon_{excess}(\infty) = \sigma_n^2 + \frac{\mu P_{update}\sigma_n^2}{(2-\mu P_{update})}\frac{Trace(E(C(k)))}{Trace(E(C(1,1)))}$$

$$(1.45)$$

Thus it can be found that $\beta = \frac{\mu P_{update}}{(2-\mu P_{update})}\frac{Trace(E(C(k)))}{Trace(E(C(1,1)))}$ then the steady-state MSE of the proposed DS-APA is found to be similar to APA when $P_{update} = 1$. In addition, the step size is multiplied by P_{update}, which indicates that it affects the rate of convergence of the algorithm.

The algorithm is given in form as pseudo code as follows:

Initialization:

Set $x(0) = 0$, $w(0) = 0$, $e(0) = 0$,

Fix μ in the range $0 < \mu \leq 1$, set P, the projection order, as some integer, set δ, the regularization, and choose P_{update} as probability of update $0 < P_{update} \leq 1$

$$\text{Obtain } \beta = \frac{\mu P_{update}}{(2-\mu P_{update})}\frac{Trace(E(C(k)))}{Trace(E(C(1,1)))} \tag{1.45}$$

$$\text{Compute } \gamma_{min} = \frac{2(1+\beta)}{|S|}1_i^{-1}\left(P_{update}1\left(\frac{|S|}{2}\right)\right). \tag{1.13}$$

Given input $x(k)$, $d(k)$
For time index $k = 1,2\ldots$
Obtain $e(n) = d(n) - y(n)$ where $y(k) = A(k)w(k)$

$$\text{Compute } \gamma_{max} = \frac{k_e}{k_G}\frac{2(1+\beta)}{|S|}1_i^{-1}\left(P_{update}1\left(\frac{|S|}{2}\right)\right) \tag{1.15}$$

where $k_e = \frac{E[e^4(k)]}{E^2[e^2(k)]}$ $k_G = 3$, $k_e = 12/|S|$

If $\|e(k)\|^2 > \gamma_{min}\sigma_n^2$ and $\|e(k)\|^2 \leq \gamma_{max}\sigma_n^2$

$$w(k+1) = w(k) + \mu A^T(k)\left(\delta I + A(k)A^T(k)\right)^{-1}e(k) \tag{1.16}$$

Else

$\quad\quad$ If $\|e(\mathrm{k})\|^2 > \gamma_{max}\sigma_n^2$

$\quad\quad e(k) = 0, \; \boldsymbol{d}(k) = 0$

$\quad\quad$ End if

$\quad \boldsymbol{w}(k + 1) = \boldsymbol{w}(k)$

$\quad$ endif

1.4.3 Simulations

In this section, we present simulation results to validate the performance improvement of the proposed scheme. The results are shown in two different scenarios, namely, for system identification and real-time speech processing applications.

1.4.4 Experimental setup

The performance of the proposed data selection scheme can be evaluated by proper selection of the various parameters inherent to the APA algorithm. The main parameters involved in the algorithms are μ and P_{update}, which should lie between 0 and 1 in order to maintain stability [6,16]. As seen from Eq. (1.45), the steady-state MSE of the proposed algorithm has μ multiplied by P_{update}. For different values of P_{update}, which ranges from 0 to 1, and for step size varying between 0 and 1, it is found that steady-state MSE is smoothly varying only when μ is selected within 0.5. Therefore μ is selected as less than 0.5 throughout the experiment.

Scenario 1

Consider an unknown system with eight filter coefficients. The problem under consideration is the unknown system identification problem. The system and the filter are assumed to have the same number of filter coefficients. The system is made of the following coefficients given by

$$[0.13 - 0.34 - 0.6\,0.9 - 0.1 - 0.77\,0.79\,0.01]$$

The input is both white Gaussian and colored. The colored input is obtained by passing white Gaussian noise through first-order AR system and fourth-order AR system with transfer function given by $\frac{1}{(1-0.8\,z^{-1})}$ and $\frac{1}{(1-0.5z^{-1}+0.75z^{-2}+0.12z^{-3}+0.45z^{-4})}$, respectively. The noise is made of both Gaussian and other noise sources such as impulsive noise. For the Gaussian noise, the variance of 0.01 is set. The impulsive noise is generated by using $k(n)A(n)$, where $k(n)$ is a Bernoulli process with probability of success $[k(n) = 1] = p_r$.

The first experiment compares the performance of the proposed data-selective APA with the DS-APA algorithm proposed in Ref. [16]. For comparison, LMS and DS-LMS are also simulated. The parameters for algorithms are $\mu = 0.2$ and $P = 2$. The probability of update is chosen to be $P_{update} = 0.35$.

In Fig. 1.3, the MSE learning curves of the data–selective LMS, DS–LMS, DS–APA, and the proposed scheme are compared for white input. As expected, the performance of the proposed scheme is similar to that of the DS–APA algorithm [16]. Similar conclusions are drawn from Figs. 1.4 and 1.5 where the same experiment is repeated for colored inputs for systems of order 1 and 4, respectively. The number of the times the update is made is found to be 35% of the total size of the data set without any compromise in performance.

The second experiment is conducted with different outliers, namely, impulsive noise sources with different values of p_r. As seen from Figs. 1.6–1.8, the proposed algorithm provides satisfactory performance for all types of outliers.

In Figs. 1.9 and 1.10, the convergence of the proposed DS-APA for different values of projection order ranging from $P = 1$ to $P = 4$ for AR1 input is shown. As expected, large values of projection order leads to faster convergence than small values of P, which is similar to conventional APA.

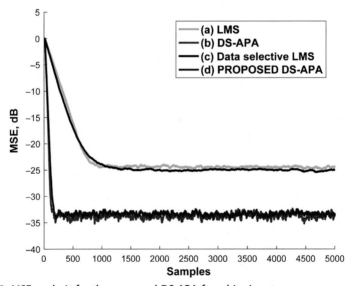

Fig. 1.3 MSE analysis for the proposed DS-APA for white input.

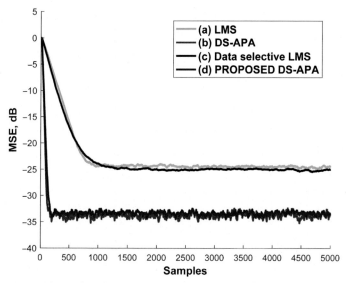

Fig. 1.4 MSE analysis of proposed DS-APA algorithm for AR1 input.

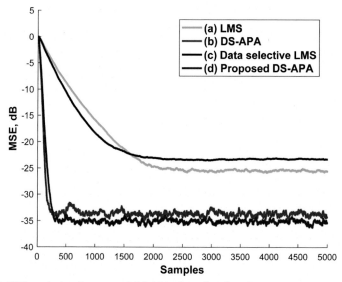

Fig. 1.5 MSE analysis of proposed DS-APA algorithm for AR4 input.

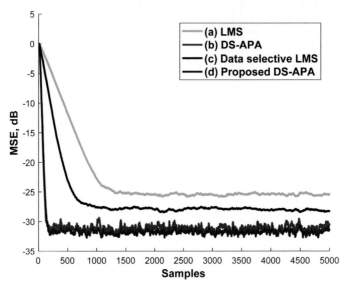

Fig. 1.6 MSE analysis of proposed DS-APA for white input with outlier as impulsive noise with $P_r = 0.01$.

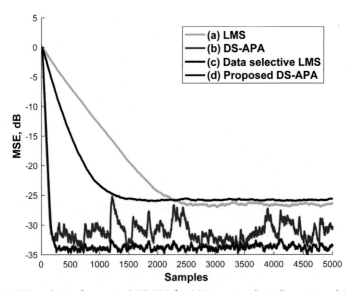

Fig. 1.7 MSE analysis of proposed DS-APA for AR1 input with outlier as impulsive noise with $P_r = 0.01$.

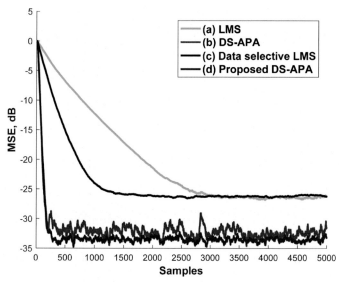

Fig. 1.8 MSE analysis of proposed DS-APA for AR4 input with outlier as impulsive noise with $P_r = 0.01$.

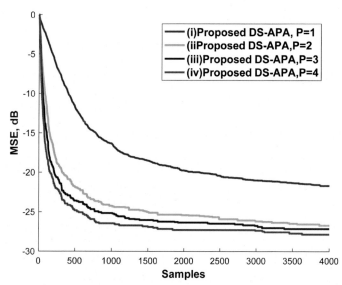

Fig. 1.9 Convergence analysis of proposed DS-APA for different projection order for white input.

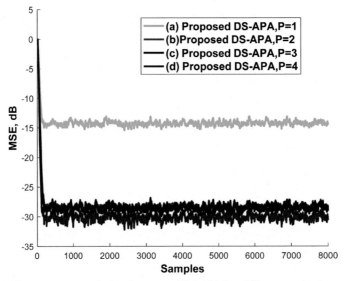

Fig. 1.10 Convergence analysis of proposed DS-APA for different projection order for AR1 input.

A similar conclusion can be obtained from Fig. 1.10 where the input is white Gaussian.

Fig. 1.11 shows the correlation between the set values of P_{update} and the simulated results obtained for the projection order set as $P = 2$ and $\mu = 0.4$, $\sigma_n^2 = 0.01$. As expected, they coincide, which proves that the proposed update rule based on Eq. (1.10) works well. Similarly in Fig. 1.12, the variation of noise and the threshold values are plotted. Thus as the noise variance is increased, the lower threshold also increases so as to update only when the variance of error is greater that the lower threshold values.

Scenario 2

In the second scenario, real-time speech data and nonuniform noise are collected and the removal of noise is done using the adaptive filter. The plot of real-time speech data and nonuniform noise are shown in Figs. 1.13 and 1.14 respectively. Here $P = 10$, $\mu = 0.4$. Fig. 1.15 illustrates the behavior of the proposed filter obtained using the proposed rule. Thus from Fig. 1.15 it is evident that the proposed rule works well for all types of noise, which indicates that it is a suitable candidate for speech processing applications. Moreover the data set used is only 45% of the total available data, which claims for the reduction of storage space, as expected.

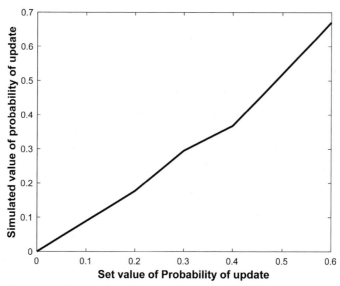

Fig. 1.11 Behavior of probability of update obtained through simulations as compared to the set value of P_{update} at $\mu = 0.4$, $\sigma_n^2 = 0.01$.

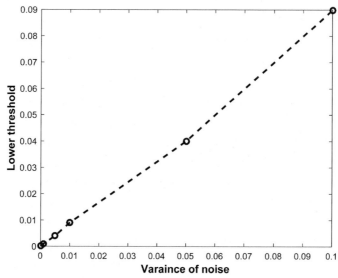

Fig. 1.12 Values of lower threshold for various values of variance of noise at $P_{update} = 0.3$, $\mu = 0.2$.

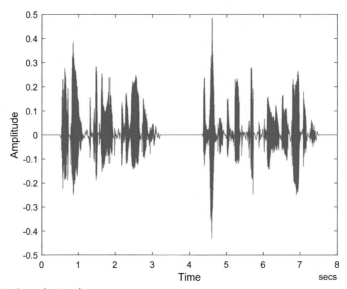

Fig. 1.13 Speech signal.

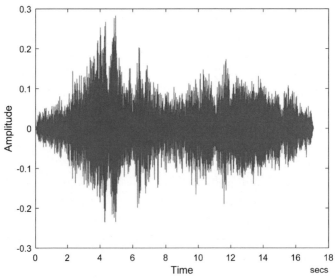

Fig. 1.14 Nonuniform noise source.

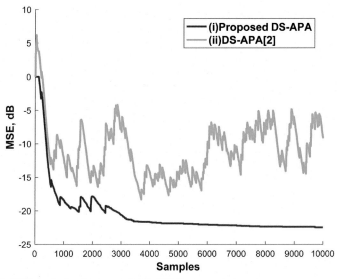

Fig. 1.15 MSE analysis of proposed DS-APA for speech input.

1.5 Discussion

The main feature of the proposed adaptive algorithm lies in the data selection criteria for the extraction of useful information from the speech signal as well as the removal of noise and outliers under different environmental conditions. Two scenarios were selected: scenario 1 is simulated and scenario 2 is the real data. The noise source is made up of both Gaussian background noise and impulsive noise as outliers with the input selected as both white and colored with the update P_{update} as 0.3. From the first and second experiments, it can be concluded that the proposed algorithm converges faster than its counterpart with lower steady-state error for all ranges of noise and outliers. In addition, it can be observed that only 35% of the total data set is used for obtaining the desired result, which can improve the speed and reduce the computational complexity of the speech processing algorithms. From Fig. 1.11 it is evident that the set value and P_{update} obtained from Eq. (1.12) matches well, proving the effectiveness of the analytical results. Next, the capability of the data selectivity of the proposed algorithm for different levels of noise was tested by plotting different noise levels and threshold values, as shown in Fig. 1.12. Thus it can be concluded that the proposed scheme satisfies the update requirement by having increased value for greater noise levels. The effect of projection order on the proposed algorithm is shown in Figs. 1.9 and 1.10. From the results it can be concluded that greater projection order improves the performance at the cost of increased complexity.

1.6 Conclusions

In this work, we developed modified data selection for APA for speech processing applications. The proposed scheme results in improved performance by the use of the proposed update rule. Further, the non-Gaussian outliers and noise are also removed by preventing the updating of coefficients during the occurrence of noise and outliers. Additionally, by the use of the probability-based update rule, reduction in computational complexity is achieved. The simulations were done to prove the analytical results obtained. Thus the proposed scheme is more suitable for speech processing applications where reduction in space and time, together with an increase in efficiency, is obtained. Further research will shed light on the application of data-selection schemes for information theory criteria-based adaptive algorithms. The issues relating to the identification of error and outliers and the optimal selection of various parameters will also be dealt with in the future.

References

[1] S. Sen, A. Dutta, N. Dey, Audio Processing and Speech Recognition: Concepts, Techniques and Research Overviews, Springer, 2019.
[2] N. Dey (Ed.), Intelligent Speech Signal Processing, Academic Press, 2019.
[3] S. Sen, A. Dutta, N. Dey, Speech processing and recognition system, in: Audio Processing and Speech Recognition, Springer, Singapore, 2019, pp. 13–43.
[4] N. Dey, A.S. Ashour, W.S. Mohamed, N.G. Nguyen, Acoustic sensors in biomedical applications, in: Acoustic Sensors for Biomedical Applications, Springer, Cham, 2019, pp. 43–47.
[5] D. Berberidis, V. Kekatos, G.B. Giannakis, Online censoring for large-scale regressions with application to streaming big data, IEEE Trans. Signal Process. 64 (15) (2016) 3854–3867.
[6] A.H. Syed, Adaptive Filter Theory, John Wily and Sons, Inc., Hoboken, NJ, 2003.
[7] C. Paleologu, et al., A variable step-size affine projection algorithm designed for acoustic echo cancellation, IEEE Trans. Audio Speech Lang. Process. 16 (8) (2008) 1466–1478.
[8] P.S. Diniz, Introduction to adaptive filtering, in: Adaptive Filtering, Springer, Cham, 2020, pp. 1–8.
[9] S.S. Haykin, Adaptive Filter Theory, Pearson Education India, 2005.
[10] M.V.S. Lima, et al., Sparsity-aware data-selective adaptive filters, IEEE Trans. Signal Process. 62 (17) (2014) 4557–4572.
[11] R.C. De Lamare, P.S.R. Diniz, Set-membership adaptive algorithms based on time-varying error bounds for CDMA interference suppression, IEEE Trans. Veh. Technol. 58 (2) (2008) 644–654.

[12] S. Zhang, J. Zhang, Set-membership NLMS algorithm with robust error bound, IEEE Trans. Circuits Syst. II Express Briefs 61 (7) (2014) 536–540.

[13] A. Zardadi, Data selection with set-membership affine projection algorithm, AIMS Electron. Electr. Eng. 3 (4) (2019) 359.

[14] P.S.R. Diniz, H. Yazdanpanah, Data censoring with set-membership algorithms, in: 2017 IEEE Global Conference on Signal and Information Processing (GlobalSIP), IEEE, 2017.

[15] Z. Wang, et al., Distributed recursive least-squares with data-adaptive censoring, in: 2017 IEEE International Conference on Acoustics, Speech and Signal Processing (ICASSP), IEEE, 2017.

[16] P.S.R. Diniz, On data-selective adaptive filtering, IEEE Trans. Signal Process. 66 (16) (2018) 4239–4252.

[17] M.J.M. Spelta, W.A. Martins, Normalized LMS algorithm and data-selective strategies for adaptive graph signal estimation, Signal Process. 167 (2020), 107326.

[18] M.O.K. Mendonça, et al., On fast converging data-selective adaptive filtering, Algorithms 12 (1) (2019) 4.

[19] C.G. Tsinos, P.S.R. Diniz, Data-selective LMS-Newton and LMS-quasi-Newton algorithms, in: ICASSP 2019-2019 IEEE International Conference on Acoustics, Speech and Signal Processing (ICASSP), IEEE, 2019.

[20] S. Radhika, A. Sivabalan, Steady-state analysis of sparsity-aware affine projection sign algorithm for impulsive environment, Circuits Syst. Signal Process. (2016) 1–14.

[21] A. Papoulis, S.U. Pillai, Probability, Random Variables, and Stochastic Processes, Tata McGraw-Hill Education, 2002.

[22] J.J. Shynk, Probability, Random Variables, and Random Processes: Theory and Signal Processing Applications, John Wiley & Sons, 2012.

[23] S. Zhao, B. Chen, J.C. Príncipe, An adaptive kernel width update for correntropy, in: The 2012 International Joint Conference on Neural Networks (IJCNN), Brisbane, QLD, 2012, pp. 1–5.

[24] O. Tanrikulu, A.G. Constantinides, Least-mean kurtosis: a novel higher-order statistics based adaptive filtering algorithm, Electron. Lett. 30 (3) (1994) 189–190.

[25] R. Wang, Y. He, C. Huang, X. Wang, W. Cao, A novel least-mean kurtosis adaptive filtering algorithm based on geometric algebra, IEEE Access 7 (2019) 78298–78310.

CHAPTER 2

Recursive noise estimation-based Wiener filtering for monaural speech enhancement

Navneet Upadhyay[a] and Hamurabi Gamboa Rosales[b]
[a]Department of Electronics and Communication Engineering, The LNM Institute of Information Technology, Jaipur, India
[b]Department of Signal Processing and Acoustics, Faculty of Electrical Engineering, Autonomous University of Zacatecas, Zacatecas, Mexico

2.1. Introduction

In practical noisy situations, clean speech signals are often contaminated by unwanted background noise from the surrounding environment. As a result speech enhancement is an important research topic in the field of signal processing. The goal of speech enhancement is to remove background noise and improve the perceptual quality and intelligibility of the speech signal, resulting in pleasant-sounding and understandable speech.

Over the decades, several monaural speech enhancement methods have been presented to eliminate noise and improve speech quality and intelligibility. Spectral subtraction concentrates on eliminating the spectral effects of acoustically added broadband noise in speech. Windowed sections of the signal are transformed to the frequency domain using fast Fourier transforms (FFTs). Estimates of the magnitude of the noise spectrum are subtracted from the signal spectrum. The enhanced speech is obtained by taking the inverse FFT (IFFT).

Boll (1979) presented a spectral subtraction method where he obtained the estimate of the noise spectrum during nonspeech periods of the input signal [1]. In the specialized case of using the system with a linear predictive coding (LPC) bandwidth compression device, he was able to obtain improvements in intelligibility. However, in the general case, the method was only successful in improving the pleasantness and inconspicuousness of the background noise; intelligibility was relatively unimproved.

Berouti (1979) noted that Boll's method had a tendency to induce a ringing or musical noise in the speech estimate [2]. He claimed this noise was

Applied Speech Processing
https://doi.org/10.1016/B978-0-12-823898-1.00007-2

© 2021 Elsevier Inc.
All rights reserved.

27

derived from the relatively large excursions in the estimated noise spectrum. He proposed two modifications to Boll's method: subtraction of an overestimate of the noise spectrum and the imposition of a minimal spectral floor beneath which the spectra components were prevented from descending. The spectral floor was intended to effectively mask the musical noise cited above. While the subjects preferred the quality of the enhanced speech, the intelligibility was the same as that of the unprocessed signal. In some noise situations, the intelligibility was worse.

The conventional spectral subtraction method is simple and obtains a noise estimate from segments of the signal where no speech is present. Nevertheless, the enhanced signal obtained by the spectral subtraction method is not optimal. Therefore, we now turn our attention to Wiener filtering, which is conceptually similar to spectral subtraction and substitutes the direct subtraction with an optimal estimate of the clean speech spectrum in a minimum mean square error (MMSE) sense [3, 4].

The chapter presents a recursive noise estimation-integrated Wiener filtering (WF-RANS) to enhance speech corrupted by background noise. First-order recursive relation, using a smoothing parameter with a value between 0 and 1, estimates the noise. The proposed approach finds the trade-off between the amount of noise reduction, speech distortion, and the level of remnant noise in a perceptive view.

The chapter is structured as follows. Section 2.2 reviews the spectral subtraction method for eliminating the estimated noise from the observed signal and describes the spectral subtraction method in the filtering domain framework. Section 2.3 outlines the recursive approach for noise estimation. Section 2.4 details the proposed WF-RANS scheme. Finally, Section 2.5 presents the objective and subjective results of the WF-RANS.

2.2. Spectral subtraction method

For the execution of the spectral subtraction method, few assumptions are important to consider. Firstly, the speech signal is assumed to quasistationary; secondly, the speech and noise should be additive and uncorrelated; and thirdly, the phase of the noisy speech is kept unchanged [1].

Since the phase information is widely considered to be unimportant in the perception of speech signals, only the noisy amplitude spectrum is processed by a spectral attenuation gain, which is derived under assumptions on the statistical characteristics of the time-frequency signals of speech and noise [1, 5, 6].

When speech signal is corrupted by additive noise, the signal that reaches the microphone can be written as $y[n] = s[n] + d[n]$, where $y[n]$ is the noisy input speech and $s[n]$ represents the clean speech signal, which is corrupted by additive background noise $d[n]$. As we know that the speech signal is non-stationary in nature and thus its processing is carried out in number of frames, the noisy speech can be expressed as

$$y[n, k] = s[n, k] + d[n, k], \quad k = 1, 2, \dots, N \tag{2.1}$$

Here n is the discrete-time index, k is the frame number and N is the total frame number.

$$Y(\omega, k) = S(\omega, k) + D(\omega, k) \tag{2.2}$$

Here ω is the discrete-angular frequency index of the frames.

$$Y(\omega, k) = \sum_{n=-\infty}^{\infty} y[n]w[k-n]e^{-j\omega n} \tag{2.3}$$

$w[n]$ is the analysis window, which is time reversed and shifted by k samples $\{w[n-k]\}$. Multiplying both sides of Eq. (2.2) by their complex conjugates, we get

$$|Y(\omega, k)|^2 = |S(\omega, k)|^2 + |D(\omega, k)|^2 + 2|S(\omega, k)D(\omega, k)| \tag{2.4}$$

where $|S(\omega,k)|^2$ is the short-term power spectrum of speech. The $|D(\omega,k)^2|$ and $|S(\omega,k)D(\omega,k)|$ terms cannot be obtained directly and are approximate as

$$E\{|Y(\omega, k)|^2\} = E\{|S(\omega, k)|^2\} + E\{|D(\omega, k)|^2\} + 2E\{|S(\omega, k)D(\omega, k)|\} \tag{2.5}$$

where $E\{\cdot\}$ denotes the expectation operator. As the additive noise is assumed to be zero-mean and uncorrelated, the $|S(\omega,k)D(\omega,k)|$ term reduces to zero, so Eq. (2.5) can be rewritten as

$$E\{|Y(\omega, k)|^2\} = E\{|S(\omega, k)|^2\} + E\{|D(\omega, k)|^2\} \tag{2.6}$$

Normally, $E\{|D(\omega, k)|^2\}$ is noise estimated during nonspeech activity periods and is denoted by $\hat{P}_d(\omega, k)$. Therefore, the estimate of the clean speech power spectrum can be written as

$$\hat{P}_s(\omega, k) \approx P_y(\omega, k) - \hat{P}_d(\omega, k) \tag{2.7}$$

Here $\hat{P}_s(\omega, k)$ is the enhanced speech-power spectrum, $P_y(\omega, k)$ is the noisy-speech power spectrum, and $\hat{P}_d(\omega, k)$ is the noise-power spectrum

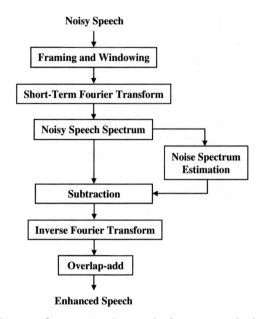

Noisy Speech

Framing and Windowing

Short-Term Fourier Transform

Noisy Speech Spectrum

Noise Spectrum Estimation

Subtraction

Inverse Fourier Transform

Overlap-add

Enhanced Speech

Fig. 2.1 Block-diagram of conventional spectral subtraction method.

that is taken from speaker silence frames [7]. The block diagram of the spectral subtraction method is given in Fig. 2.1.

The main shortcoming of the spectral subtraction method is that it produces an annoying noise with a musical character in the enhanced speech [1, 5–7]. Several variations of the spectral subtraction method have been documented to overcome the musical noise [1, 5–7].

The modified form of Eq. (2.7) is

$$\hat{P}_s(\omega, k) = P_y(\omega, k)\left[1 - \frac{\hat{P}_d(\omega, k)}{P_y(\omega, k)}\right]$$
$$= P_y(\omega, k)H(\omega) \tag{2.8}$$

Here, $H(\omega)$ is the filter gain, which is a real value.

2.3. Recursive noise estimation

Noise reduction from noisy speech is the most crucial step of monaural speech enhancement methods because the quality of the enhanced speech depends on the accurate estimation of the noise-power spectrum. A commonly used method for noise spectrum estimation is to average oversections in the input speech signal that do not contain speech. However, this approach requires that nonspeech sections can be detected reliably, which is difficult, especially under noisy conditions. Moreover, it relies on the fact that there exists a sufficient amount of nonspeech in the signal [8]. In order to avoid these problems, we propose an approach to estimate the noise spectrum without explicit frame-wise speech/nonspeech classification.

In our approach, the noise is estimated by averaging past spectral power values, using a smoothing parameter. The first-order recursive relation for noise power estimation is

$$\hat{P}_d(\omega, k) = \alpha \hat{P}_d(\omega, k-1) + (1-\alpha)P_y(\omega, k) \tag{2.9}$$

where α ($0 < \alpha < 1$) is a smoothing parameter, k is the current frame-index, ω is the frequency bin-index, $P_y(\omega, k)$ is the short-term power spectrum of noisy speech, $\hat{P}_d(\omega, k)$ is the noise-power spectrum estimate in ωth frequency bin of current frame, and $\hat{P}_d(\omega, k-1)$ is the past noise-power spectrum estimate.

2.4. Recursive noise estimation-based Wiener filtering

The Wiener filter is the optimum filter with respect to minimizing the mean square error (MMSE) between the filter outputs and the desired clean speech signal. It gives an MMSE estimate of the short-term Fourier transform (STFT), whereas the spectral subtraction obtains an MMSE estimate of the short-term spectral magnitude without affecting the phase [3, 4, 9]. Thus, the Wiener filter in the frequency domain is

$$H(\omega) = \frac{P_s(\omega)}{P_y(\omega)} = \frac{P_s(\omega)}{P_s(\omega) + P_d(\omega)} \tag{2.10}$$

Here $P_s(\omega)$ and $P_d(\omega)$ are the signal and noise-power spectrum, respectively.

$$H_{Wiener}(\omega) = \frac{P_s(\omega)}{P_y(\omega)} = \frac{P_y(\omega) - P_d(\omega)}{P_y(\omega)} \tag{2.11}$$

The estimated speech-power spectrum, from Eq. (2.11), is

$$\hat{P}_s(\omega, k) = H_{Wiener}(\omega)P_y(\omega, k) \tag{2.12}$$

The estimated speech signal can be obtained by taking the inverse FFT of Eq. (2.12)

$$|\hat{s}[n]| = IFFT\left[\sqrt{\hat{P}_s(\omega, k)}\right] \tag{2.13}$$

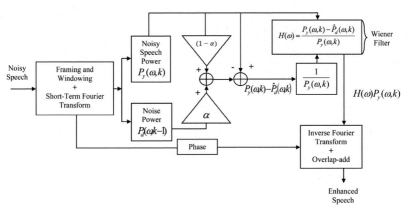

Fig. 2.2 Block-diagram of recursive noise estimation-based Wiener filtering (WF-RANS).

The estimate of the clean speech spectral magnitude combined with the noisy phase gives an estimate of the enhanced speech signal as

$$\hat{s}[n] = |\hat{s}[n]| \angle y[n] \tag{2.14}$$

The block diagram of the WF-RANS is given in Fig. 2.2.

2.5. Experimental setup and results

In this section, we outline the experimental results and performance evaluation of the proposed WF-RANS and compare WF-RANS with conventional spectral subtraction (SS) method. We took clean speech samples (sampled at 8 kHz) from the NOIZEUS corpus speech database [10] and used four different utterances (two male speakers and two female speakers) for the experimental work.

The background noises have different time-frequency distributions and have different impacts on the speech signal. Therefore, we have taken computer-generated white Gaussian noise (AWGN) and pink noise for the evaluation of our speech enhancement approach. For the AWGN, we have taken noisy speech samples from the NOIZEUS corpus while pink noise is mixed with four speech sentences at different SNR levels to prepare the speech database [10].

In experiments, the noise samples used are of zero–mean, and the energy of the noisy speech samples are normalized to unity. The frame size is taken to be 200 samples with 50% overlapping. The Hamming window with 200 samples is applied to each frame individually. The windowed speech frame is then analyzed using FFT with a length of 256 samples. The noise is estimated from the noisy speech using Eq. (2.9).

For the performance evaluations of the WF-RANS approach, we used the objective evaluations of global SNR, SNRseg, and PESQ score, and for subjective evaluations, we used speech spectrograms with informal hearing tests [11–13].

From the extensive study, it is observed that for input SNR (in the range of 0–15 dB), the value of smoothing parameter α increases and the value of SNRseg is better for both AWGN and pink noise. This is also validated by the PESQ score, which is the objective measure of subjective speech quality.

As described in Eq. (2.9), the noise estimation approach in the current frame is heavily dependent on noise in the previous frame as well as lightly dependent on noisy speech in the current frame. Therefore, as shown in Tables 2.1 and 2.2, $\alpha = 0.8$ is the suitable value for the WF-RANS approach.

Table 2.3 lists the objective results of SNR, SNRseg, and PESQ score for AWGN and pink noises. The overall SNR and SNRseg values are better for the WF-RANS scheme in comparison to the spectral subtraction method for both AWGN and pink noise at varying SNR levels. In addition, the PESQ score is higher for pink noise in comparison to AWGN noise.

Furthermore, it is noticed from the SNR point of view that the speech enhancement for a female speaker is better than for a male speaker for both AWGN and pink noise cases. From the SNRseg point of view, the speech enhancement for the female speaker is better than for the male speaker for AWGN noise cases. Nevertheless, it is almost the same as the male speaker for the pink noise case. Therefore, the WF-RANS scheme outperforms for the female speaker compared to the male speaker and even better in the pink noise case.

Usually, a speech enhancement scheme produces two main undesirable effects: musical noise and speech distortion. However, these effects are difficult to quantify with objective measures. Therefore, a speech spectrogram, which is a visual representation to analyze the time-frequency distribution of speech, is used in our work for observing the structure of musical noise and speech distortion. As shown in Figs. 2.3–2.10, the WF-RANS approach reduces musical noise more than the spectral subtraction method does. Therefore, enhanced speech by the proposed approach (WF-RANS) is more pleasing and the musical character has a *perceptual quality* while distortion leftovers are tolerable. This is verified by the global SNR, SNRseg, and PESQ score, and validated by a speech spectrogram with informal hearing tests.

Table 2.1 SNRseg and PESQ score of recursive noise estimation-based Wiener filtering for AWGN noise at varying SNR.

SNR (dB)		SNRseg of recursive noise estimation-based Wiener filtering for AWGN noise at varying SNR									PESQ score at α = 0.8 and α = 0.9	
		α = 0.1	α = 0.2	α = 0.3	α = 0.4	α = 0.5	α = 0.6	α = 0.7	α = 0.8	α = 0.9	α = 0.8	α = 0.9
0	M1	6.1467	6.0659	6.0133	6.0399	5.9901	6.0175	6.0454	6.0616	6.1187	1.6303	1.1772
5		6.3677	6.2702	6.2695	6.2913	6.3015	6.3206	6.3580	6.4312	6.5102	2.0182	1.6995
10		6.5676	6.5376	6.5385	6.5256	6.5806	6.5890	6.6768	6.7096	6.8621	2.2176	2.0143
15		6.8252	6.7900	6.7987	6.7972	6.8166	6.8741	6.9444	6.9868	7.1393	2.1856	2.0428
0	M2	6.2204	6.1461	6.0805	6.0686	6.0728	6.0423	6.0459	6.0922	6.0729	1.1713	1.0589
5		6.3976	6.3580	6.3465	6.3312	6.3470	6.3469	6.3834	6.4227	6.4842	1.6756	1.6657
10		6.6059	6.5684	6.5739	6.5787	6.5155	6.6381	6.6834	6.7625	6.8366	2.1777	2.1624
15		6.8084	6.7695	6.7817	6.8107	6.8579	6.9055	6.9370	6.9716	7.0725	2.3571	2.3358
0	F1	5.9614	5.7786	5.6670	5.5736	5.4823	5.4748	5.4169	5.4051	5.4607	0.9820	1.0797
5		6.2681	6.1205	6.0240	5.9785	5.9385	5.9351	5.9432	5.9614	6.0864	1.9816	1.7841
10		6.5369	6.4735	6.4010	6.4322	6.3949	6.3962	6.4655	6.4918	6.5995	2.3012	2.3000
15		6.7649	6.7180	6.7246	6.7218	6.7190	6.7818	6.7973	6.8897	6.9835	2.3712	2.3602
0	F2	6.4262	6.3111	6.2797	6.2683	6.2425	6.2454	6.2461	6.2627	6.2864	1.2494	1.1830
5		6.6165	6.5597	6.5419	6.5284	6.5407	6.5734	6.5902	6.6821	6.7447	1.8662	1.8383
10		6.8300	6.8104	6.8302	6.7961	6.8220	6.8639	6.9143	6.9915	7.0786	2.2525	2.2472
15		6.9911	7.0216	7.0443	7.0635	7.0785	7.1195	7.1900	7.2309	7.3318	2.3947	2.1773

Note: Bold values indicate that if α = 0.8 the improvement is more than other value of α.

Table 2.2 SNRseg and PESQ score of recursive noise estimation-based Wiener filtering for Pink noise at varying SNR.

SNR (dB)		SNRseg of recursive noise estimation-based Wiener filtering for Pink noise at varying SNR									PESQ score at α = 0.8 and α = 0.9	
		α = 0.1	α = 0.2	α = 0.3	α = 0.4	α = 0.5	α = 0.6	α = 0.7	α = 0.8	α = 0.9	α = 0.8	α = 0.9
0	M1	7.1804	7.6478	8.1224	8.6437	9.2580	10.0242	10.9384	12.1556	14.1045	2.2797	2.2561
5		7.5343	8.0645	8.6094	9.2089	9.9063	10.7526	11.8025	13.1289	15.1599	2.2900	2.2246
10		7.8623	8.4459	9.0351	9.6901	10.4369	11.3619	12.4804	13.8679	15.9073	2.3506	2.3502
15		8.1451	8.7907	9.4050	10.1007	10.8864	11.8497	12.9939	14.4112	16.4543	2.3128	2.3722
0	M2	7.4840	8.0469	8.6217	9.2468	9.9331	10.7268	11.7171	12.9963	14.4753	2.3540	2.3487
5		7.8239	8.4938	9.1513	9.8675	10.6583	11.5395	12.5952	13.9591	15.4976	2.4514	2.4250
10		8.1056	8.8379	9.5594	10.3449	11.1972	12.1281	13.2258	14.6550	16.2260	2.5167	2.4523
15		8.3597	9.1274	9.8854	10.6985	11.5778	12.5605	13.6931	15.1609	16.7699	2.4551	2.4262
0	F1	7.1277	7.5197	7.9687	8.5086	9.1902	10.0361	11.0740	12.2320	13.5888	2.4532	2.5074
5		7.6803	8.2478	8.8196	9.4581	10.2177	11.1460	12.2515	13.3636	14.6915	2.5948	2.5671
10		8.0851	8.7913	9.4689	10.1752	10.9703	11.9247	13.0274	14.0961	15.4108	2.5813	2.5442
15		8.3932	9.1846	9.9260	10.6615	11.4682	12.4157	13.5044	14.5681	15.8823	2.5331	2.5087
0	F2	7.6561	8.1930	8.7320	9.2916	9.9363	10.7461	11.7180	12.8056	14.0104	2.4605	2.4499
5		8.0090	8.6549	9.2737	9.9188	10.6028	11.4735	12.5068	13.6579	14.9412	2.4708	2.4466
10		8.2927	9.0171	9.7005	10.4007	11.1334	12.0158	13.0696	14.2717	15.6377	2.3723	2.3414
15		8.5195	9.3201	10.055	10.7813	11.5283	12.4203	13.4850	14.7086	16.1560	2.3317	2.2999

Note: Bold values indicate that if α = 0.8 the improvement is more than other value of α.

Table 2.3 Output SNR, Output SNRseg and Perceptual evaluation of speech quality (PESQ) measure results of enhanced speech signals at (0, 5, 10, 15) dB SNRs. English sentence "The line where the edges join was clean," produced by M1 speaker, "The sky that morning was clear and bright blue," produced by M2 speaker, "The set of china hit the floor with a crash," produced by F1 speaker and "She has a smart way of wearing clothes," produced by F2 speaker is used as original signal.

Noise type		Enhancement approach	SNR (dB)				SNRseg (dB)				PESQ score			
			0 dB	5 dB	10 dB	15 dB	0 dB	5 dB	10 dB	15 dB	0 dB	5 dB	10 dB	15 dB
M1	AWGN	SS	1.415	2.177	2.942	4.214	6.066	6.440	6.724	6.986	1.0502	1.6001	1.8948	1.9426
		WF-RANS	1.441	2.210	2.944	4.245	6.079	6.444	6.765	7.025	1.6404	2.0182	2.2176	2.1856
	Pink	WF-RANS	4.248	4.442	4.482	4.407	12.555	14.128	14.867	14.411	2.2197	2.2900	2.4506	2.4128
M2	AWGN	SS	2.105	4.542	4.504	5.046	6.072	6.449	6.776	7.001	1.0011	1.6595	2.1124	2.1925
		WF-RANS	2.056	4.462	4.460	5.042	6.067	6.422	6.774	6.991	0.9414	1.6756	2.1177	2.2571
	Pink	WF-RANS	5.174	5.296	5.446	5.449	12.996	14.956	14.655	15.160	2.4540	2.4514	2.5167	2.4551
F1	AWGN	SS	2.289	4.844	5.011	5.564	5.426	5.981	6.497	6.849	0.9950	1.6768	2.1407	2.4589
		WF-RANS	2.405	4.900	4.964	5.641	5.448	6.000	6.557	6.877	0.9820	1.6816	2.2012	2.4712
	Pink	WF-RANS	5.750	5.885	5.940	5.944	12.242	14.464	14.096	14.568	2.4542	2.5948	2.5814	2.5441
F2	AWGN	SS	1.548	2.709	4.418	4.900	6.250	6.654	6.977	7.244	1.1425	1.8454	2.0407	2.2450
		WF-RANS	1.776	2.795	4.416	4.944	6.264	6.674	6.991	7.245	1.1494	1.8062	2.1425	2.4947
	Pink	WF-RANS	4.978	4.074	4.107	4.115	12.805	14.657	14.271	14.708	2.4605	2.4708	2.4724	2.4417

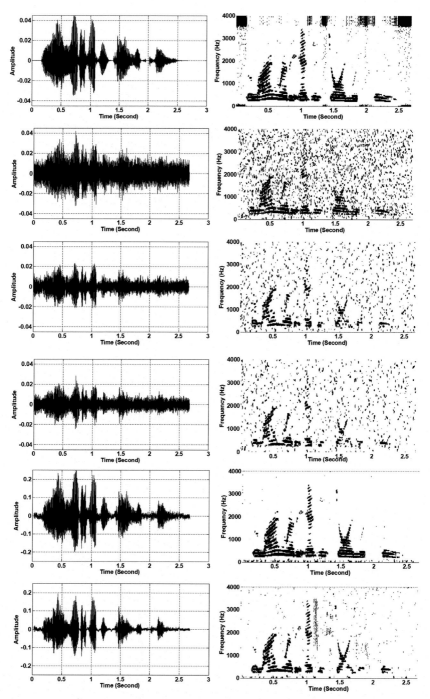

Fig. 2.3 Waveforms and spectrograms (from top to bottom): (i) clean speech: sp22 utterances, "The line where the edges join was clean" by M1 speaker; (ii) noisy speech (White noise at 0 dB); (iii) enhanced speech by SS; (iv) enhanced speech by WF-RANS; (v) speech degraded by Pink noise (0 dB SNR); (vi) enhanced speech by WF-RANS.

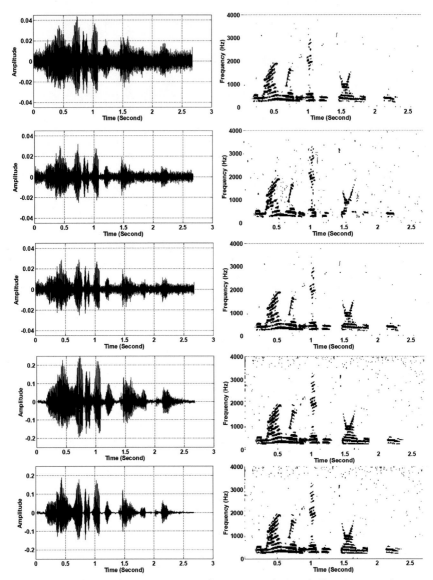

Fig. 2.4 Waveforms and spectrograms (from top to bottom): (i) noisy speech: sp22 utterances, "The line where the edges join was clean" by M1 speaker degraded by AWGN noise (5 dB SNR); (ii) enhanced speech by SS; (iii) enhanced speech by WF-RANS; (iv) speech degraded by Pink noise (5 dB SNR); (v) enhanced speech by WF-RANS.

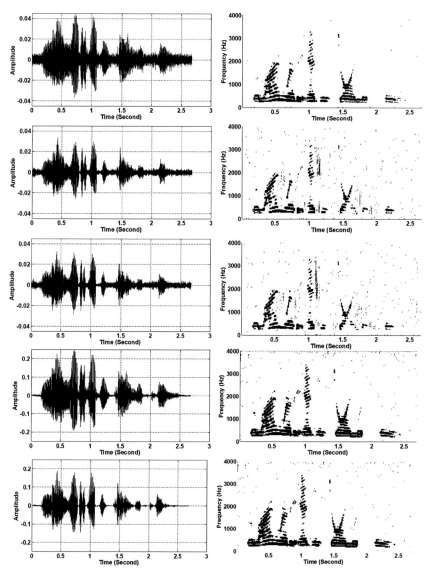

Fig. 2.5 Waveforms and spectrograms (from top to bottom): (i) noisy speech: sp22 utterances, "The line where the edges join was clean" by M1 speaker degraded by AWGN noise (10 dB SNR); (ii) enhanced speech by SS; (iii) enhanced speech by WF-RANS; (iv) speech degraded by Pink noise (10 dB SNR); (v) enhanced speech by WF-RANS.

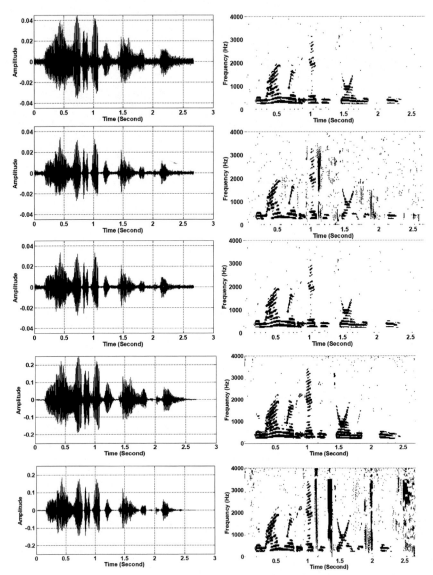

Fig. 2.6 Waveforms and spectrograms (from top to bottom): (i) noisy speech: sp22 utterances, "The line where the edges join was clean" by M1 speaker degraded by AWGN noise (15 dB SNR); (ii) enhanced speech by SS; (iii) enhanced speech by WF-RANS; (iv) speech degraded by Pink noise (15 dB SNR); (v) enhanced speech by WF-RANS.

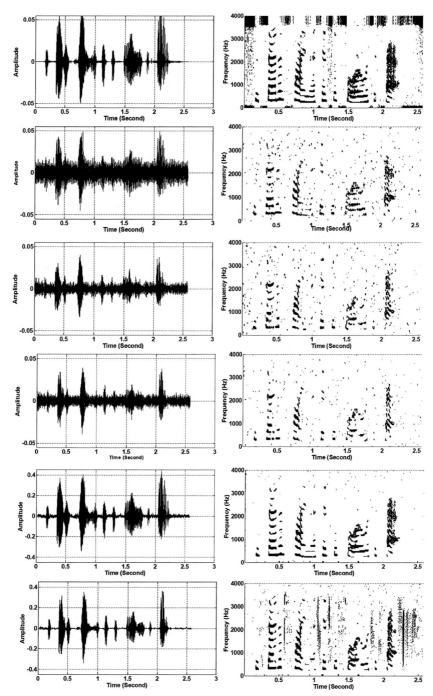

Fig. 2.7 Waveforms and spectrograms (from top to bottom): (i) clean speech: sp20 utterances, "The set of china hit the floor with a crash" by F1 speaker; (ii) noisy speech (White noise at 0 dB); (iii) enhanced speech by SS; (iv) enhanced speech by WF-RANS (v) degraded speech: sp26 utterances, "The set of china hit the floor with a crash" by F1 speaker by Pink noise (5 dB SNR); (vi) enhanced speech by WF-RANS.

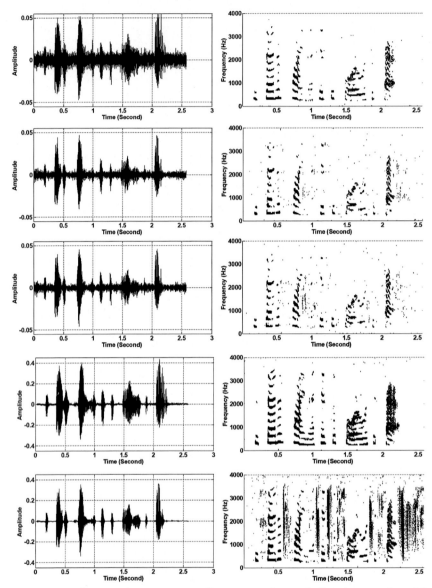

Fig. 2.8 Waveforms and spectrograms (from top to bottom): (i) noisy speech: sp20 utterances, "The set of china hit the floor with a crash" by F1 speaker degraded by AWGN noise (5 dB SNR); (ii) enhanced speech by SS; (iii) enhanced speech by WF-RANS (iv) degraded speech: sp26 utterances, "The set of china hit the floor with a crash" by F1 speaker by Pink noise (5 dB SNR); (v) enhanced speech by WF-RANS.

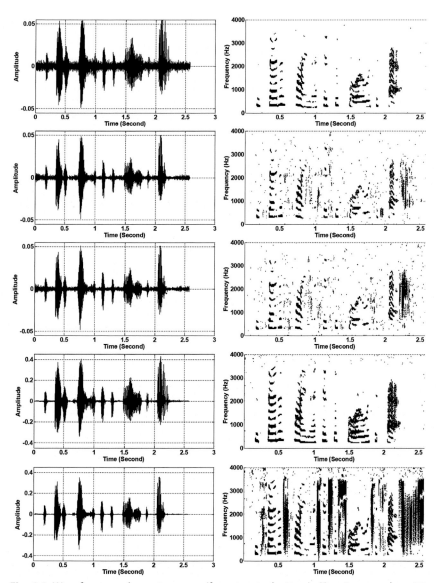

Fig. 2.9 Waveforms and spectrograms (from top to bottom): (i) noisy speech: sp20 utterances, "The set of china hit the floor with a crash" by F1 speaker degraded by AWGN noise (10 dB SNR); (ii) enhanced speech by SS; (iii) enhanced speech by WF-RANS; (iv) degraded speech: sp20 utterances, "The set of china hit the floor with a crash" by F1 speaker by Pink noise (10 dB SNR); (v) enhanced speech by WF-RANS.

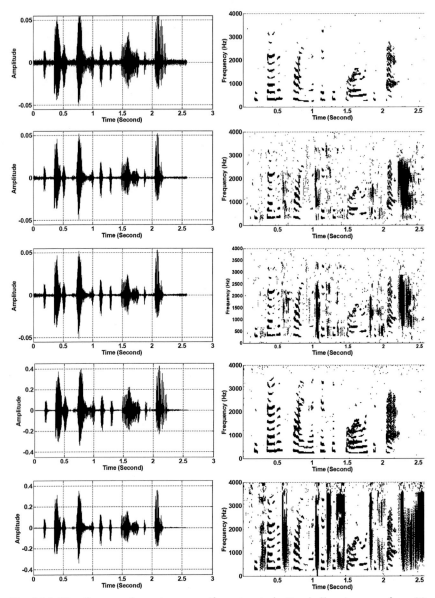

Fig. 2.10 Waveforms and spectrograms (from top to bottom): (i) noisy speech: sp20 utterances, "The set of china hit the floor with a crash" by F1 speaker degraded by AWGN noise (15 dB SNR); (ii) enhanced speech by SS; (iii) enhanced speech by WF-RANS; (iv) degraded speech: sp20 utterances, "The set of china hit the floor with a crash" by F1 speaker by Pink noise (15 dB SNR); (v) enhanced speech by WF-RANS.

2.6. Conclusion

In this chapter, we investigated the WF-RANS approach to enhance monaural speech. We used both objective and subjective measures at varying SNR levels for AWGN and pink noise and compared WF-RANS with conventional spectral subtraction methods. The obtained results show that the quality of speech enhanced by WF-RANS is good and the musical noise is less structured than in conventional methods, while the distortion of speech remains acceptable.

References

[1] S.F. Boll, Suppression of acoustic noise in speech using spectral subtraction, IEEE Trans. Acoust. Speech Signal Process. 27 (2) (1979) 113–120.

[2] M. Berouti, R. Schwartz, J. Makhoul, Enhancement of speech corrupted by acoustic noise, in: Proceedings of Int. Conf. on Acoustics, Speech, and Signal Processing, Washington, DC, April 1979, , pp. 208–211.

[3] B. Xia, C. Bao, Wiener filtering based speech enhancement with weighted denoising auto-encoder and noise classification, Speech Commun. 60 (2014) 13–29.

[4] F. El, M.A. Abd, Speech enhancement with an adaptive Wiener filter, Int. J. Speech Technol. 17 (1) (2014) 53–64.

[5] P.C. Loizou, Speech Enhancement: Theory and Practice, second ed., CRC Press, 2013.

[6] S.V. Vaseghi, Advanced Digital Signal Processing and Noise Reduction, third ed., Wiley, 2006.

[7] N. Upadhyay, A. Karmakar, The spectral subtractive-type algorithms for enhancing speech in noisy environments, in: IEEE International Conference on Recent Advances in Information Technology, IIT(ISM) Dhanbad, India, March, 2012, , pp. 841–847.

[8] M. Rainer, Noise power spectral density estimation based on optimal smoothing and minimum statistics, IEEE Trans. Speech Audio Process. 9 (5) (2001) 504–512.

[9] S. Haykin, Adaptive Filter Theory, fourth ed., Prentice Hall, Upper Saddle River, 2003.

[10] A Noisy Speech Corpus for Assessment of Speech Enhancement Algorithms, http://ecs.utdallas.edu/loizou/speech/noizeus/ and http://www.speech.cs.cmu.edu/.

[11] Y. Hu, P.C. Loizou, Evaluation of objective quality measures for speech enhancement, IEEE Trans. Audio Speech Lang. Process. 16 (1) (Jan. 2008) 229–238.

[12] Perceptual Evaluation of Speech Quality (PESQ): An Objective Method for End-to-End Speech Quality Assessment of Narrowband Telephone Networks and Speech Codecs, ITU-T Recommendation P.862.1, (2003).

[13] H. Yi, P.C. Loizou, Subjective comparison of speech enhancement algorithms, in: IEEE International Conference on Acoustics, Speech and Signal Processing, Toulouse, France, May 14–19, 2006.

CHAPTER 3

Modified least mean square adaptive filter for speech enhancement

M. Kalamani[a] and M. Krishnamoorthi[b]

[a]Department of Electronics and Communication Engineering, Velalar College of Engineering and Technology, Erode, India
[b]Department of Computer Science and Engineering, Dr. N.G.P. Institute of Technology, Coimbatore, India

3.1. Introduction

Speech is fundamental to human communication. Speech signals convey much more information than spoken words. The information conveyed by speech is multilayered and includes time and frequency modulation of such carriers of information as formants, pitch, and intonation. Speech perception refers to the process in which humans are able to interpret and understand the sounds used in language.

The study of speech perception is closely connected to the fields of phonetics, phonology, linguistics, cognitive psychology, and perception. Research on speech perception seeks to know how human listeners recognize speech sounds and use this information to understand spoken language. Speech perception is articulated in terms of speech quality and speech intelligibility. Speech quality is the overall impression of the listener to say how good the quality of the speech is. Since, the natural air-transmitted speech which is emitted from a real human every day are heard and this speech provides a reference point on the quality scale. The listeners rate the speech under test relative to this reference [1, 2].

Speech intelligibility is the accuracy with which one can hear what is being said. It is measured as the percentage of correctly identified responses relative to the number of responses. The relationship between perceived quality and speech intelligibility is not entirely understood. However, there exists some correlation between the two. Generally, speech perceived as good quality gives high intelligibility, and vice versa. However, there are samples that are rated as poor quality but that give high intelligibility scores, and vice versa [3].

Applied Speech Processing
https://doi.org/10.1016/B978-0-12-823898-1.00004-7
© 2021 Elsevier Inc.
All rights reserved.
47

Background noise is added with the desired speech signal in most of the speech processing applications form an additive mixtures which are picked up by microphone. It has no correlation with the desired speech signal and it can be stationary or nonstationary, white or colored. This additive background noise causes speech degradation and often occurs due to sources such as air conditioning units, fans, cars, city streets, factory environments, helicopters, computer systems, and so on. In speech processing applications like mobile phones, hands-free phones, car communication, teleconference systems, hearing aids, voice coders, automatic speech recognition, and forensics, the major challenge is to eliminate background noise. Speech enhancement algorithms are widely used for these applications in order to remove noise from degraded speech in a noisy environment. Speech enhancement algorithms aim to ameliorate both the quality and intelligibility of the speech signal by reducing background noise. Most of the speech enhancement algorithms are based on the analysis-modification-synthesis framework. Some of the existing noise reduction methods include spectral subtraction, Wiener filter, adaptive filter, and minimum mean square error short time spectral amplitude (MMSE-STSA) estimation methods [4–6].

Speech recognition is an active area of research area, especially regarding interacting with computers through speech by the disabled community. Under various noisy environments, large vocabulary continuous speech recognition (LVCSR) research investigates different languages because each language has its own specific features. In the last few decades, the need for a continuous speech recognition (CSR) system in the Tamil language has greatly increased. In this application, the clean speech signal is corrupted by background noise. A major challenge in this system is to eliminate the noise from the noisy signal. In order to improve human perception, the speech enhancement techniques are extensively used as front-end processing in the CSR system under noisy environments.

Many solutions have been developed to deal with the problem of noisy speech enhancement. However, the developed algorithms introduce speech distortion and residual noise under nonstationary noisy environments. Residual noise occurs due to noise reduction, the noise signal is present in the enhanced clean speech signal during the speech pause period, and speech distortion is due to the part of the speech spectrum lost during the noise reduction process. Therefore, it has been found that the noise reduction process is more effective for improving speech quality and affects the intelligibility of the clean speech signal.

The main objective of this chapter is to enhance the noisy Tamil speech signal with acceptable speech quality and intelligibility using the modified least mean square adaptive noise reduction (LMS-ANR) algorithm in order to reduce speech distortion and residual noise.

3.2. Literature survey

3.2.1 Introduction

Noise reduction approaches improve signal quality with acceptable intelligibility by reducing background additive noise. Some types of noise reduction methods include optimum filters, adaptive filters, and spectral subtraction methods [6]. Cornelis et al. [7] initially proposed optimum filters for background noise reduction. This filter requires prior information about the signals used for processing. In addition, this filter's performance depends on signal spectrum, which introduces signal suppression in magnitude without altering the phase spectrum.

3.2.2 Speech enhancement using adaptive filters

Recently the adaptive filter was developed to overcome the problems introduced by optimum filters for noise reduction. The least mean square (LMS) adaptive filter is a simple and iterative-based noise reduction algorithm that reduces the mean square error (MSE). This adaptive algorithm does not require any prior information about the signals considered for processing [8]. Chi et al. [9] described a filtered-x LMS (FxLMS) filter for adaptive noise control applications to reduce the secondary path effect. This algorithm has fast convergence and produced good sound quality for speech signals. However, this filter produce more MSE [10]. Rahman et al. [11] proposed a Block LMS (BLMS) algorithm for noise reduction. In this approach, for each block of data, only one times the filter coefficients are updated. Hence, this technique introduces an MSE with less computational requirements. Huang and Lee [12] described the normalized LMS (NLMS) algorithm for noise reduction in order to resolve the problem of fast convergence with less MSE. In this algorithm, estimated noise spectrum and MSE control changes in step size. This approach has a fast convergence rate with simple implementation. However, this filter produces more noise and signal distortion [13].

3.3. Optimum filter for noise reduction

3.3.1 Speech signal modeling

The sampled version of a band-limited noisy signal is expressed as the sum of a clean speech signal $s(n)$ and a background noise $d(n)$, which is described as follows:

$$x(n) = s(n) + d(n) \qquad (3.1)$$

where "n" indicates the sample index. Let us assume the noisy speech signal occurred as a nonstationary process and statistically independent with zero mean.

The noisy speech signal as divided into frames with "L" samples using windows. Then, fast Fourier transform (FFT) is computed on these frames. Further, computation of subsequent FFT is carried out by shifting windows by "R" samples. The Fourier transform of the noisy speech signal can be demonstrated as

$$X(\lambda, k) = \sum_{n=0}^{L-1} x(\lambda R + n) h(n) e^{-j\Omega_k n} \qquad (3.2)$$

where λ is the frame index of the subsample, k is the frequency bin of each frame that lies $k \in \{0, 1, ..., L-1\}$ and the normalized mid frequency Ω_k is given by $\Omega_k = \frac{2\pi k}{L}$. Then the additive noisy model of the speech signal is described as

$$X(\lambda, k) = S(\lambda, k) + D(\lambda, k) \qquad (3.3)$$

where $X(\lambda, k)$, $S(\lambda, k)$, and $D(\lambda, k)$ are the noisy, clean, and noise signal STFT coefficients evaluated for each frame, respectively. The magnitude of all three signals is computed for further performance analysis. For each k, the noise spectral variance is given by $\sigma_D^2 = E(|D(\lambda, k)|^2))$ and the speech spectral variance as $\sigma_S^2 = E(|S(\lambda, k)|^2)$. A priori signal-to-noise ratio (SNR) is defined by

$$\xi(\lambda, k) = \frac{\sigma_S^2(\lambda, k)}{\sigma_D^2(\lambda, k)} \qquad (3.4)$$

and a posteriori SNR is expressed as follows [14]

$$\gamma(\lambda, k) = \frac{X^2(\lambda, k)}{\sigma_D^2(\lambda, k)} \qquad (3.5)$$

3.3.2 Noise reduction using Wiener filter

The Wiener filter is an optimal and linear time-invariant filter used to enhance signal intelligibility and quality of speech by reducing additive noise in the background. Consider the clean speech signal and noise signal are stationary random process and uncorrelated to each other. This filter reduces the MSE to an acceptable level, but introduces signal distortion in the enhanced speech signal [15].

The purpose of this optimum filter design is to produce an enhanced speech signal with less MSE. Let us consider the noisy signal as $x(n)$ described in Eq. (3.1). In this, the clean speech signal and noise are statistically independent stationary random processes.

Noise reduction using the Wiener filter is demonstrated in Fig. 3.1. In this, the enhanced signal produced at the output of this process is described as

$$e(n) = x(n) - \hat{d}(n) \tag{3.6}$$

where $\hat{d}(n)$ is the response of this optimum filter and is described as

$$y(n) = \hat{d}(n) = \sum_{i=0}^{N-1} w(i)d(n-i) \tag{3.7}$$

where $w(i)$ is the filter coefficients of $(N-1)$th order and is calculated to minimize MSE. Here, consider $d(n)$ as having zero mean characteristics and being uncorrelated with $s(n)$. Then, the Wiener Hopf equation becomes

$$[R_s(n) + R_d(n)]w(n) = r_s(n) \tag{3.8}$$

Hence, the filter coefficients $w(n)$ is evaluated as follows

$$w(n) = [R_s(n) + R_d(n)]^{-1} r_s(n) \tag{3.9}$$

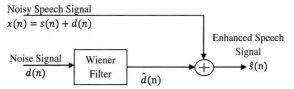

Fig. 3.1 Block diagram for noise reduction using Wiener filter.

3.4. Noise reduction using least mean square adaptive algorithms

3.4.1 Introduction

The signal available in the environment is always nonstationary. For these conditions, the Wiener filter fails to enhance the speech signal. To overcome these issues during noise reduction, the adaptive filter has been developed. These filter coefficients are automatically adapted whenever changes occur in the input signal [8].

Fig. 3.2 shows the noise reduction process using adaptive filters. This filter requires prior information about the noise signal. Due to random characteristics of the signal and noise, the filter has time-varying characteristics. Hence, the design of this adaptive filter is more complex than that of the optimum filter. In this, the coefficients of the filter are updated automatically at each time "n" in order to produce less MSE. It is described as follows

$$w_{n+1} = w_n + \Delta w_n \qquad (3.10)$$

where Δw_n is the correction term added to w_n at time "n" to get a different set of filter coefficients w_{n+1} at time "$n+1$." Here, the signal statistics are unknown due to nonstationary conditions. This filter enhances the speech signal from its signal characteristics.

The filter uses the iterative procedure such as steepest decent approach to minimize the MSE. The weight is updated based on ensemble averages and is described as

$$w_{n+1} = w_n + \mu E[e(n)d(n)] \qquad (3.11)$$

where μ is the step size, which controls the convergence speed of the filter. In most applications, these expected values $E[e(n)d(n)]$ are unknown and obtained from the signal characteristics.

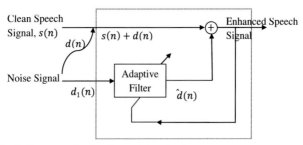

Fig. 3.2 Block diagram of adaptive noise reduction.

3.4.2 LMS adaptive filter

For less computational complexity with minimum MSE, the conventional LMS adaptive filter is introduced for noise elimination. In this filter, the estimation of signal statistics is replaced by the sample mean. For this, the filter coefficients are updated as follows

$$w_{n+1} = w_n + \mu e(n)d(n) \tag{3.12}$$

where $d(n)$ is the reference input noise signal and response $y(n)$ of the adaptive filter is described as

$$y(n) = w_n^T d(n) \tag{3.13}$$

and μ is the step size, which controls the speed of convergence. The error signal $e(n)$ can be generated by filter output $y(n)$ and the noisy signal $x(n)$ is given by

$$e(n) = x(n) - y(n) \equiv \hat{s}(n) \tag{3.14}$$

where $\hat{s}(n)$ is the response of this process and is considered as enhanced clean speech signal [16,17].

The LMS adaptive filter is the most famous and frequently used efficient adaptive algorithm for noise reduction. Performance of this filter depends on the filter order, signal nature, and step size (μ). Further, the convergence criterion is described as

$$0 < \mu < \frac{2}{\varphi_{\max}} \tag{3.15}$$

where $\varphi_{\max}$ is the maximum Eigen value in correlation matrix R_x of the speech signal [18].

3.4.3 Block LMS adaptive filter

In the block LMS (BLMS), the coefficients of the filter are constant for each block of the signal. Using these coefficients, the response of this filter $y(n)$ and its error signal $e(n)$ are calculated over for each block. At the end of each block of signal, the coefficient of this filter is updated using L gradient estimates.

This filter is used to split the speech signal $x(n)$ into nonoverlapping blocks of length L and considers one block for processing at a time. After receiving each block of speech samples, the filter weights are updated. For kth block, the response of the filter is described as

$$y(kL + l) = w_{kL}^T d(kL + l) \tag{3.16}$$

and the error signal is given by

$$e(kL + l) = x(kL + l) - y(kL + l) \qquad (3.17)$$

where L is the block length and $x(n)$ is the noisy signal. The weight of the kth block is updated as

$$w_{(k+1)L} = w_{kL} + \mu \frac{1}{L} \sum_{l=0}^{L-1} e(kL + l) d(kL + l) \qquad (3.18)$$

where μ is the step size and minimizes the MSE. Due to block processing, this filter has less computational complexity and introduces more speech distortion [11,19].

3.4.4 Filtered-X LMS adaptive filter

In this adaptive filter approach, the input speech signal is filtered before it is processed in a standard LMS algorithm. Hence, it reduces secondary path effects [20]. Further, the coefficients of this filter are updated as

$$w_{n+1} = w_n + \mu e(n) d'(n) \qquad (3.19)$$

where μ is the convergence parameter and $e(n)$ is the error signal described as

$$e(n) = x(n) - y(n) \qquad (3.20)$$

where $x(n)$ is the noisy signal and $y(n)$ is the output of the adaptive filter, which is given as

$$y(n) = w_n^T \cdot d(n) \qquad (3.21)$$

Then, $d'(n)$ is the response of the filter and is described as

$$d'(n) = C_n^T d(n) \qquad (3.22)$$

where $d(n)$ is the reference noise signal and C_n is the coefficients of the filter and is updated as follows

$$C_{n+1} = C_n + \mu e'(n) y(n) \qquad (3.23)$$

where $e'(n)$ the error signal occurred during filtering and is described as

$$e'(n) = e(n) - r(n) \qquad (3.24)$$

where $r_i(n)$ is response of the filter and is described as

$$r_i(n) = C_n^T \cdot y(n) \qquad (3.25)$$

In this, the convergence speed slows down due to large order [9,21,22].

3.4.5 Normalized LMS adaptive filter

The adaptive filters discussed earlier have slow convergence due to statistically dependent step-size parameter. Generally, in the nonstationary noise environment, the statistics of speech signals are unknown [12]. In order to minimize MSE, the convergence condition for step size is represented as

$$0 < \mu < \frac{2}{N E\{|d(n)|^2\}} \tag{3.26}$$

where N is the order of the filter and $E\{|d(n)|^2\}$ is the power of signal $d(n)$, which is estimated as follows

$$\hat{E}\{|d(n)|^2\} = \frac{1}{N} \sum_{k=0}^{N-1} |d(n-k)|^2 \tag{3.27}$$

Eq. (3.26) is modified as follows

$$0 < \mu < \frac{2}{d^H(n) \cdot d(n)} \tag{3.28}$$

In this NLMS method, the normalized step size is expressed as

$$\mu(n) = \frac{\beta}{d^H(n) \cdot d(n)} = \frac{\beta}{\|d(n)\|^2} \tag{3.29}$$

where β is the normalized step size with $0 < \beta < 2$. In this filter, the weights are updated as

$$w_{n+1} = w_n + \beta \frac{d(n)}{\|d(n)\|^2} e(n) \tag{3.30}$$

NLMS has fast convergence speed when compared to standard LMS adaptive algorithms [13]. In order to overcome noise amplification under small or large input signal conditions, this filter's coefficients are modified as follows

$$w_{n+1} = w_n + \beta \frac{d(n)}{\epsilon + \|d(n)\|^2} e(n) \tag{3.31}$$

where ϵ is small positive integer [23,24].

3.4.6 Noise reduction using modified least mean square adaptive filter (LMS-ANR)

In the NLMS adaptive filter, the estimation of norm value under nonstationary noise environment is a tedious process. Hence, it will affect the amplitude of enhanced speech signal, which creates more speech distortion and leads to more MSE. In order to reduce MSE, the proposed adaptive filter is developed. The convergence parameter of this filter is described as follows

$$\mu(n) = \frac{2\alpha}{N(\beta + d(n)d^T(n))} \quad \text{for } 0 < \alpha < 1 \text{ and } 0 < \beta < 2 \quad (3.32)$$

where N is order of the filter; $d(n)$ is reference noise signal, α is smoothing factor; and β is normalized step-size. Appropriate choice of these filter parameters results in fast convergence with less MSE.

In this chapter, convergence parameter of the filter μ depends on the signal characteristics, smoothing factor, filter order, and normalized step size. Best possible values of these parameters are calculated easily with less computational requirement and they are fixed as stable values depending on the signal conditions.

Then, the filter coefficients w_{n+1} are updated from the previous instance of filter coefficients w_n in time domain and it is described as

$$w_{n+1} = w_n + \frac{2\alpha d(n)}{N(\beta + d(n)d^T(n))} e(n) \quad (3.33)$$

The various steps for noise reduction using the proposed adaptive filter are clearly demonstrated in the following algorithm as follows:

Algorithm 3.1 Noise reduction using modified least mean square adaptive filter

1: **For all** values of sample index "n"
2: Compute the response of proposed and existing filter $y(n)$ through Eq. (3.13)
3: Then, calculate the error signal $e(n)$ using Eq. (3.14) under various noise environments
4: Further estimate the convergence parameter $\mu(n)$ as it is described in Eq. (3.32) by considering the optimal values as $N = 5$, $\alpha = 0.95$, and $\beta = 0.001$ obtained through simulation
5: Finally, update the filter coefficients w_{n+1} in time domain as expressed in Eq. (3.33)
6: **end** for

3.5. Experimental results and discussions

3.5.1 Dataset description and experimental setup

Throughout the evaluation process, the different input speech signals of different speakers are obtained from the database of International Institute of Information Technology (IIIT) consisting of various Tamil speech sentences. Around 400–500 different sentences from each speaker (nine speakers) are used for evaluation. In addition, the different input SNR levels (0, 5, 10, and 15 dB) are considered along with various input noise signals obtained from the noisy speech corpus (NOIZEUS) database consisting of a variety of environmental conditions such as airport, car, babble, exhibition, restaurant, street, station, and train noises. In this work, the noise signals are extracted from the NOIZEUS database and added to the speech signal of the IIIT database using SPDemo software for the evaluation of various adaptive noise reduction algorithms.

All the adaptive noise reduction algorithms discussed in this chapter are simulated using MATLAB R2017a software. In this chapter, the results are validated through various performance metrics like peak signal-to-noise ratio (PSNR), mean opinion score (MOS), segmental SNR improvement (ΔSNR_{seg}), mean square error (MSE), and log spectral distance (LSD).

3.5.2 Performance comparisons of adaptive noise reduction algorithms

This section presents the performance metrics of the proposed LMS–ANR for enhancing the noisy speech signal. The algorithm is simulated with different speech sentences from the two databases under various noise environments as previously mentioned. Objective and subjective performance measures are evaluated and compared between the proposed and existing adaptive filters. In the proposed adaptive filter, the convergence parameter μ is attained from the observation of signal and does not require the signal statistical values. Fig. 3.3 shows the average value of convergence parameter (μ) for each frame using NLMS–ANR and the proposed LMS–ANR algorithms. Results show that there is an improvement of μ value in each frame for the proposed adaptive filter as compared with the existing filter. This performance improvement provides faster convergence with less MSE.

Fig. 3.4 illustrates the performance evaluation of PSNR in dB for sp01 sentence of NOIZEUS and tam_0010 sentence of IIIT database under all noise environments mentioned earlier. Here, there is an improvement in PSNR value while the input speech SNR levels vary from 0 to 15 dB

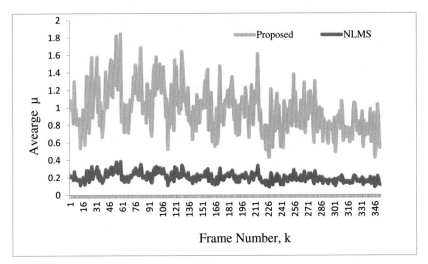

Fig. 3.3 Average value of convergence parameter (μ) for each frame using NLMS and proposed adaptive filters.

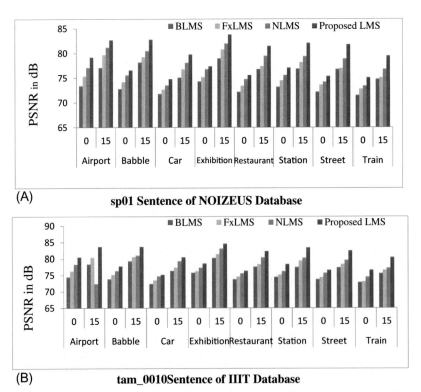

Fig. 3.4 PSNR in dB for the proposed LMS-ANR and existing LMS algorithms for different database speech sentences under various noise environments with different input SNR levels (0–15 dB).

due to reduction in noise signal levels. Results show PSNR improvement of 0.6–2.9 dB in sp01 sentence and 0.5–3.3 dB in tam_0010 sentence for the proposed adaptive filter when compared to the other adaptive filters. MOS is the most used subjective measure and is applied to validate speech signal quality. The listeners can provide their rating in five different integer scales. This experiment is carried out for 40 listeners from different educational backgrounds. The experiment is randomly tested with clean, noisy, and enhanced speech signals 10 times. Then, the rating is allotted from the listeners and MOS is evaluated by averaging the rating of all the listeners.

Tables 3.1 and 3.2 depict the SegSNR improvement (Δ SNR$_{seg}$) in dB, log spectral distance (LSD), mean opinion score (MOS) and MSE for sp01 sentence under different noise environment for existing and proposed adaptive filters. Results indicate that there is a reduction in Δ SNR$_{seg}$ and an improvement in MOS values when the input SNR level of the noise signal is increased from 0 to 15 dB under various noises. In addition, results show that there is a drop in LSD and MSE values when there is an improvement in noise input SNR levels. Tables 3.3 and 3.4 evaluate and compare performance measures between all adaptive algorithms for tam_0010 sentence of the IIIT database.

The proposed LMS-ANR filter produces an improvement of Δ SNR$_{seg}$ from 25.44% to 60.22% and MOS from 7.37% to 47.44% as compared with other LMS algorithms for speech sentences from different databases. In addition, there is a significant reduction in LSD from 2.05% to 40.86% and MSE from 17.65% to 59.52% for tam_0010 sentence. All the mentioned adaptive filters are validated by the time domain under different signal and noise environments. Fig. 3.5 demonstrates the time domain plot of sp01 sentence for airport noise with 5 dB input SNR level.

Fig. 3.6 shows a time domain plot of train noise with 10 dB input SNR levels. It is observed from the graphical results that the proposed adaptive filter provides a significant reduction in noise level for different NOIZEUS speech signals and produces an improved speech signal that is identical to the clean signal related to all the existing algorithms. Fig. 3.7 shows the experimental results of the tam_0010 speech sentence from the IIIT database with 0 dB car noise. Fig. 3.8 shows results for 15 dB exhibition noise. The results show that the proposed adaptive filter yields an improvement of speech quality by significantly reducing background noise level as compared to existing adaptive filters.

From the analysis of simulated results presented in the previous sections, it is concluded that the proposed LMS-ANR filter provides considerable enhancement under various noise environments as compared with other

Table 3.1 Performance comparison of MOS and ΔSNR_{seg} under different noise environments for existing and proposed adaptive filters (sp01 sentence).

Types of noise	Input SNR (dB)	BLMS		FxLMS		NLMS		Proposed LMS-ANR	
		ΔSNR_{seg} (dB)	MOS	ΔSNR_{seg} (dB)	MOS	ΔSNR_{seg} (dB)	MOS	ΔSNR_{seg} (dB)	MOS
Airport	0	−5.14	1.83	−5.06	1.95	−4.92	2.08	−4.12	2.15
	5	−5.63	2.18	−5.47	2.34	−5.28	2.52	−4.86	2.76
	10	−5.97	2.76	−5.72	2.89	−5.54	2.94	−5.19	3.07
	15	−6.35	3.09	−6.19	3.16	−5.89	3.31	−5.56	3.42
Babble	0	−5.27	1.62	−5.12	1.70	−4.86	1.78	−4.35	1.87
	5	−5.86	2.13	−5.73	2.25	−5.53	2.32	−5.17	2.40
	10	−6.13	2.63	−6.01	2.73	−5.85	2.92	−5.68	3.12
	15	−6.97	3.02	−6.82	3.17	−6.48	3.28	−6.13	3.36
Car	0	−5.31	1.87	−5.17	1.93	−5.01	2.05	−4.91	2.14
	5	−6.02	2.17	−5.92	2.29	−5.63	2.33	−5.48	2.47
	10	−6.75	2.62	−6.59	2.73	−6.25	2.96	−6.18	3.01
	15	−7.26	3.00	−7.13	3.23	−7.04	3.31	−6.86	3.43
Exhibition	0	−5.75	1.68	−5.63	1.84	−5.43	1.92	−5.28	2.04
	5	−6.26	1.93	−6.14	2.07	−5.98	2.27	−5.66	2.39
	10	−6.93	2.27	−6.81	2.42	−6.63	2.61	−6.38	2.78
	15	−7.82	2.89	−7.63	2.98	−7.44	3.02	−7.13	3.27
Restaurant	0	−5.12	2.05	−5.02	2.21	−4.87	2.36	−4.63	2.43
	5	−5.78	2.36	−5.59	2.53	−5.38	2.64	−5.16	2.76
	10	−6.12	2.72	−6.03	2.86	−5.87	2.93	−5.59	2.99
	15	−6.57	2.98	−6.40	3.03	−6.17	3.12	−6.02	3.26

Station	0	−5.65	2.18	−5.48	2.24	−5.17	2.32	−5.03	2.43
	5	−6.37	2.47	−6.12	2.61	−5.93	2.78	−5.71	2.88
	10	−7.05	2.66	−6.90	2.73	−6.69	2.89	−6.50	2.97
	15	−7.68	2.89	−7.43	2.96	−7.28	3.10	−7.13	3.26
Street	0	−5.49	1.85	−5.28	1.94	−5.12	2.03	−5.03	2.17
	5	−6.07	2.07	−5.89	2.28	−5.63	2.48	−5.46	2.63
	10	−6.86	2.48	−6.63	2.56	−6.52	2.73	−6.23	2.89
	15	−7.39	2.79	−7.13	2.90	−7.01	3.07	−6.81	3.28
Train	0	−5.28	1.75	−5.06	1.94	−4.83	2.24	−4.36	2.65
	5	−5.61	2.01	−5.43	2.28	−5.27	2.49	−4.93	2.83
	10	−5.98	2.37	−5.72	2.56	−5.54	2.73	−5.28	2.97
	15	−6.37	2.69	−6.15	2.82	−5.91	3.01	−5.56	3.34

Table 3.2 Performance comparison of LSD and MSE under different noise environments for existing and proposed adaptive filters (sp01 sentence).

Types of noise	Input SNR (dB)	BLMS		FxLMS		NLMS		Proposed LMS-ANR	
		LSD	MSE	LSD	MSE	LSD	MSE	LSD	MSE
Airport	0	7.21	0.44	7.09	0.41	6.48	0.42	6.13	0.36
	5	6.79	0.41	6.61	0.37	6.17	0.35	5.87	0.31
	10	6.27	0.35	6.18	0.32	5.89	0.31	5.54	0.25
	15	5.80	0.32	5.72	0.28	5.28	0.26	5.01	0.22
Babble	0	7.32	0.48	7.11	0.42	6.88	0.39	6.63	0.35
	5	7.10	0.43	6.89	0.38	6.54	0.34	6.38	0.32
	10	6.59	0.36	6.33	0.32	6.13	0.28	6.02	0.26
	15	5.44	0.29	5.18	0.26	5.05	0.23	4.85	0.21
Car	0	9.03	0.43	8.23	0.38	7.83	0.34	7.61	0.32
	5	8.42	0.39	8.02	0.35	7.56	0.31	7.28	0.29
	10	7.53	0.31	7.18	0.24	7.07	0.22	6.93	0.20
	15	6.12	0.28	5.93	0.21	5.80	0.19	5.52	0.16
Exhibition	0	7.60	0.39	7.42	0.36	7.36	0.34	7.18	0.31
	5	6.37	0.32	6.18	0.28	6.03	0.25	5.90	0.22
	10	5.00	0.29	4.86	0.27	4.73	0.22	4.58	0.19
	15	4.18	0.26	3.95	0.23	3.68	0.18	3.49	0.15
Restaurant	0	8.12	0.38	8.00	0.35	7.82	0.32	7.56	0.30
	5	7.56	0.33	7.37	0.31	7.18	0.29	6.89	0.25
	10	6.53	0.30	6.36	0.26	6.09	0.24	5.49	0.21
	15	5.18	0.28	4.89	0.23	4.53	0.20	4.12	0.17

Station	0	8.39	0.41	8.21	0.37	8.11	0.33	8.03	0.31
	5	7.18	0.35	7.03	0.32	6.83	0.29	6.69	0.27
	10	5.86	0.31	5.68	0.28	5.47	0.26	5.28	0.24
	15	5.02	0.26	4.96	0.24	4.38	0.22	4.18	0.19
Street	0	9.04	0.51	8.79	0.47	8.54	0.43	8.17	0.41
	5	8.38	0.46	8.12	0.42	8.00	0.40	7.89	0.37
	10	7.48	0.40	7.28	0.39	7.12	0.36	7.03	0.32
	15	6.32	0.37	6.18	0.31	6.03	0.28	5.68	0.25
Train	0	8.82	0.56	8.63	0.51	8.59	0.45	8.31	0.43
	5	8.31	0.52	8.06	0.43	7.93	0.39	7.57	0.37
	10	7.53	0.48	7.34	0.37	7.12	0.35	7.04	0.31
	15	6.87	0.45	6.59	0.33	6.18	0.31	5.92	0.24

Table 3.3 Performance comparison of MOS **and** ΔSNR_{seg} under different noise environments for existing and proposed adaptive filters (tam_0010 sentence).

Types of noise	Input SNR (dB)	BLMS		FxLMS		NLMS		Proposed LMS-ANR	
		ΔSNR_{seg} (dB)	MOS	ΔSNR_{seg} (dB)	MOS	ΔSNR_{seg} (dB)	MOS	ΔSNR_{seg} (dB)	MOS
Airport	0	-6.49	2.21	-5.62	2.52	-4.48	2.73	-3.23	3.02
	5	-7.20	2.42	-6.95	2.68	-6.36	2.88	-3.86	3.29
	10	-7.98	2.51	-7.68	2.77	-7.04	2.94	-4.13	3.41
	15	-8.06	2.72	-7.87	2.91	-7.38	3.11	-4.69	3.62
Babble	0	-6.46	2.15	-5.92	2.43	-5.07	2.78	-2.57	3.17
	5	-6.90	2.39	-6.22	2.66	-5.79	2.97	-3.05	3.35
	10	-7.09	2.55	-6.73	2.98	-6.28	3.20	-3.82	3.56
	15	-7.85	2.92	-7.44	3.10	-7.12	3.39	-4.09	3.64
Car	0	-7.46	2.03	-7.02	2.12	-6.73	2.32	-4.12	2.94
	5	-8.11	2.37	-7.93	2.50	-7.48	2.71	-4.98	3.09
	10	-8.68	2.55	-8.24	2.78	-7.99	2.90	-5.73	3.27
	15	-9.25	2.81	-8.93	2.91	-8.56	3.08	-6.04	3.41
Exhibition	0	-7.18	2.01	-6.84	2.14	-6.37	2.28	-4.02	2.72
	5	-7.96	2.18	-7.59	2.37	-7.03	2.52	-4.74	2.89
	10	-8.65	2.34	-8.14	2.59	-7.87	2.73	-5.38	3.04
	15	-8.82	2.51	-8.55	2.68	-8.11	2.85	-5.96	3.16
Restaurant	0	-6.07	2.21	-5.73	2.36	-5.02	2.53	-2.92	3.08
	5	-6.68	2.33	-6.27	2.58	-5.93	2.76	-3.74	3.25
	10	-7.12	2.56	-6.85	2.81	-6.55	2.99	-4.21	3.43
	15	-7.69	2.79	-7.40	3.04	-7.16	3.16	-4.93	3.57

Station	0	−6.88	2.02	−6.28	2.14	−5.72	2.28	−3.22	2.87
	5	−7.07	2.19	−6.73	2.46	−6.27	2.64	−3.90	2.95
	10	−7.54	2.44	−7.18	2.69	−6.99	2.86	−4.68	3.12
	15	−7.93	2.68	−7.59	2.92	−7.26	3.02	−5.27	3.28
Street	0	−6.64	2.09	−6.27	2.17	−5.92	2.38	−3.79	2.91
	5	−7.49	2.23	−7.08	2.52	−6.83	2.77	−4.52	3.05
	10	−7.90	2.48	−7.58	2.79	−7.27	2.90	−5.28	3.26
	15	−8.58	2.65	−8.26	3.00	−7.94	3.13	−5.92	3.40
Train	0	−6.41	2.00	−6.05	2.15	−5.73	2.28	−3.44	2.75
	5	−7.17	2.11	−6.85	2.26	−6.49	2.46	−4.07	2.91
	10	−7.93	2.39	−7.58	2.68	−7.03	2.81	−4.82	3.08
	15	−8.55	2.56	−8.22	2.75	−7.96	2.99	−5.58	3.34

Table 3.4 Performance comparison of LSD and MSE under different noise environments for existing and proposed adaptive filters (tam_0010 sentence).

Types of noise	Input SNR (dB)	BLMS		FxLMS		NLMS		Proposed LMS-ANR	
		LSD	MSE	LSD	MSE	LSD	MSE	LSD	MSE
Airport	0	7.87	0.58	7.15	0.52	6.58	0.46	5.12	0.31
	5	7.28	0.54	6.56	0.48	5.97	0.41	4.96	0.27
	10	6.88	0.48	5.79	0.41	5.25	0.37	4.21	0.22
	15	6.15	0.42	5.26	0.37	4.87	0.33	3.93	0.17
Babble	0	7.63	0.52	7.02	0.46	6.26	0.40	5.01	0.28
	5	7.22	0.47	6.58	0.42	5.93	0.36	4.85	0.24
	10	6.95	0.44	6.37	0.36	5.28	0.31	4.11	0.21
	15	6.12	0.39	5.86	0.31	4.85	0.27	3.78	0.18
Car	0	7.87	0.61	7.10	0.56	6.72	0.51	5.66	0.42
	5	7.59	0.57	6.82	0.52	6.33	0.46	5.21	0.37
	10	7.26	0.53	6.47	0.48	5.98	0.42	4.93	0.32
	15	6.87	0.49	6.08	0.42	5.29	0.37	4.62	0.28
Exhibition	0	7.88	0.71	7.26	0.66	6.51	0.61	5.43	0.48
	5	7.64	0.67	7.03	0.62	6.30	0.55	5.11	0.42
	10	7.48	0.63	6.93	0.57	6.02	0.52	4.94	0.36
	15	7.15	0.59	6.46	0.52	5.79	0.48	4.62	0.33
Restaurant	0	7.95	0.54	7.46	0.48	7.02	0.43	5.95	0.34
	5	7.74	0.49	7.13	0.43	6.71	0.39	5.74	0.30
	10	7.58	0.45	7.02	0.40	6.15	0.35	5.26	0.27
	15	7.14	0.41	6.81	0.37	5.99	0.30	5.01	0.22

Table 3.4 Performance comparison of LSD and MSE under different noise environments for existing and proposed adaptive filters (tam_0010 sentence)—cont'd

Types of noise	Input SNR (dB)	BLMS		FxLMS		NLMS		Proposed LMS-ANR	
		LSD	MSE	LSD	MSE	LSD	MSE	LSD	MSE
Station	0	8.02	0.64	7.63	0.58	7.28	0.53	6.04	0.41
	5	7.86	0.59	7.28	0.54	6.92	0.49	5.83	0.36
	10	7.56	0.53	7.10	0.48	6.63	0.44	5.59	0.33
	15	6.70	0.49	5.93	0.45	5.37	0.40	5.26	0.29
Street	0	7.90	0.66	7.56	0.61	7.25	0.55	6.17	0.43
	5	7.71	0.62	7.26	0.58	7.01	0.51	5.96	0.39
	10	7.48	0.59	7.03	0.53	6.72	0.48	5.67	0.35
	15	7.06	0.52	6.85	0.48	6.38	0.43	5.41	0.31
Train	0	7.97	0.59	7.63	0.53	7.22	0.48	6.09	0.39
	5	7.55	0.54	7.28	0.49	6.91	0.45	5.88	0.36
	10	7.24	0.51	7.04	0.46	6.63	0.42	5.53	0.32
	15	7.00	0.48	6.72	0.42	6.24	0.39	5.16	0.29

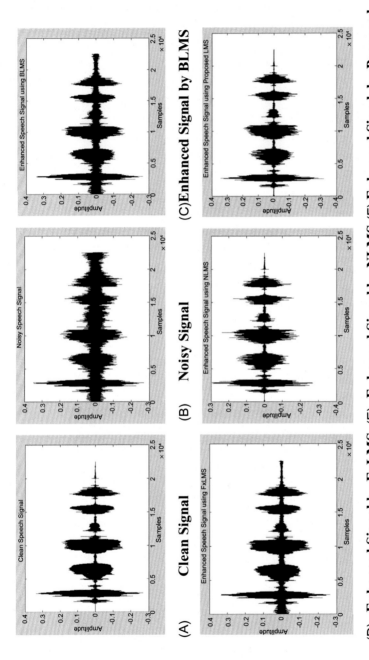

(A) **Clean Signal** (B) **Noisy Signal** (C) **Enhanced Signal by BLMS**

(D) **Enhanced Signal by FxLMS** (E) **Enhanced Signal by NLMS** (F) **Enhanced Signal by Proposed**

Fig. 3.5 Time domain representation of clean, noisy, and enhanced speech signals using existing and proposed adaptive filters for airport noise with 5 dB input SNR level (A–F—sp01 sentence).

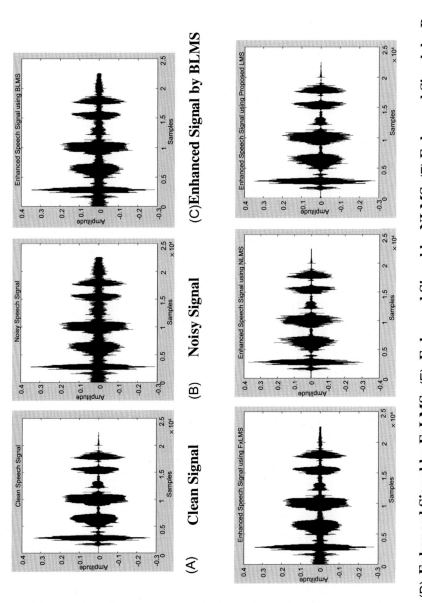

(A) **Clean Signal**   (B)   **Noisy Signal**   (C)**Enhanced Signal by BLMS**

(D) **Enhanced Signal by FxLMS**   (E) **Enhanced Signal by NLMS**   (F) **Enhanced Signal by Proposed**

Fig. 3.6 Time domain representation of clean, noisy, and enhanced speech signals using existing and proposed adaptive filters for train noise with 10 dB input SNR level (A–F—sp01 sentence).

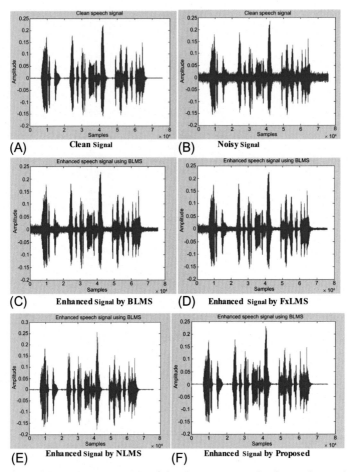

Fig. 3.7 Time domain representation of the clean, noisy, and enhanced speech signals using existing and proposed adaptive filters for the car noise with 0 dB input SNR level (A–F: tam_0010Sentence).

LMS adaptive filters. Table 3.5 shows the percentage improvement of various performances of the proposed adaptive filter compared with existing filters. This enhancement indicates that the proposed adaptive filter improves speech signal quality under different noise environments. It also minimizes speech signal distortion and residual noises at a significant level.

This adaptive noise reduction algorithm requires the noise signal as the reference signal for speech enhancement. However, for most speech processing applications, the noise information is unknown and has a nonstationary nature. Under these situations, the noise reduction algorithm using adaptive filter lags behind in enhancing the signal.

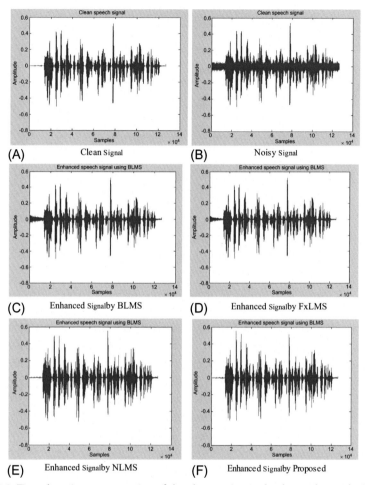

(A)   Clean Signal
(B)   Noisy Signal
(C)   Enhanced Signalby BLMS
(D)   Enhanced Signalby FxLMS
(E)   Enhanced Signalby NLMS
(F)   Enhanced Signalby Proposed

Fig. 3.8 Time domain representation of the clean, noisy, and enhanced speech signals using existing and proposed adaptive filters for the exhibition noise with 15 dB input SNR level (A–F: tam_0010Sentence).

3.6. Conclusion

In this chapter, we described noise reduction using a modified LMS adaptive filter under different noisy environments. We conducted the experiments using the NOIZEUS and IIIT Hyderabad databases for noise reduction using adaptive filters. The foremost objective of our proposed adaptive filter is to improve speech signal quality by reducing MSE value.

First, we conducted the various experiments to evaluate objective and subjective measures. Second, we used these same measures to validate the

Table 3.5 Percentage improvement of the proposed and existing adaptive filters under different noise conditions.

Database	Performance metrics	% of Improvement		
		BLMS	FxLMS	NLMS
NOIZEUS	PSNR (dB)	7.94	6.23	3.73
	Δ SNR$_{seg}$ (dB)	19.84	18.58	16.26
	MOS	51.43	36.60	18.30
	LSD	20.46	15.75	9.85
	MSE	46.67	34.78	22.58
IIIT	PSNR (dB)	8.05	5.50	4.27
	Δ SNR$_{seg}$ (dB)	60.22	56.59	49.31
	MOS	47.44	38.68	26.72
	LSD	40.86	35.49	22.19
	MSE	59.52	57.89	48.48

performance quality of the existing and proposed adaptive filters. While comparing the proposed LMS-ANR with existing algorithms, we found improvement in PSNR (up to 8.05%), ΔSNRseg (up to 60.22%), and MOS (up to 51.43%) values. In addition, the proposed adaptive filter reduced LSD value up to 40.86% and MSE value up to 59.52% as compared with existing LMS adaptive filters. From the obtained experimental results, it is found that the proposed adaptive filter significantly outperforms other adaptive filters. LMS adaptive noise reduction-based speech enhancement algorithms are developed to enhance clean speech from noisy speech and reference noise signals.

This algorithm requires noise information as the reference input signal. However, only a noisy signal is generally available. Under noisy environments, the LMS-ANR algorithms are unsuccessful for enhancing speech signals. In the future, this problem can be overcome by spectral subtraction with noise estimation algorithms for speech enhancement under nonstationary noisy environments.

References

[1] N. Dey, Intelligent Speech Signal Processing, Academic Press, 2019.
[2] N. Dey, A.S. Ashour, W.S. Mohamed, N.G. Nguyen, Acoustic sensors in biomedical applications, in: Acoustic Sensors for Biomedical Applications, Springer, Cham, 2019, pp. 43–47.

[3] P.C. Loizou, G. Kim, Reasons why current speech-enhancement algorithms do not improve speech intelligibility and suggested solutions, IEEE Trans. Audio Speech Lang. Process. 19 (1) (2011) 47–56.

[4] S. Sen, A. Dutta, N. Dey, Audio Processing and Speech Recognition: Concepts, Techniques and Research Overviews, Springer, 2019.

[5] S. Sen, A. Dutta, N. Dey, Speech Processing and Recognition System' in Audio Processing and Speech Recognition, Springer, Singapore, 2019, pp. 13–43.

[6] P. Scalart, Speech enhancement based on a priori signal to noise estimation, Proc. IEEE Int. Conf. Acoust. Speech Signal Process. 2 (1996) 629–632.

[7] B. Cornelis, M. Moonen, J. Wouters, Performance analysis of multichannel Wiener filter-based noise reduction in hearing aids under second order statistics estimation errors, IEEE Trans. Audio Speech Lang. Process. 19 (5) (2011) 1368–1381.

[8] S. Haykin, B. Widrow, Least-Mean-Square Adaptive Filters, John Wiley & Sons, 2003.

[9] H.F. Chi, S.X. Gao, S.D. Soli, A. Alwan, Band-limited feedback cancellation with a modified filtered-X LMS algorithm for hearing aids, Speech Commun. 39 (1) (2003) 147–161.

[10] J. Hellgren, Analysis of feedback cancellation in hearing aids with filtered-X LMS and the direct method of closed loop identification, IEEE Trans. Speech Audio Process. 10 (2) (2002) 119–131.

[11] M.Z.U. Rahman, R.A. Shaik, D.V. Reddy, Adaptive noise removal in the ECG using the block LMS algorithm, in: Proceedings of the Second IEEE International Conference on Adaptive Science and Technology, 2009, pp. 380–383.

[12] H.C. Huang, J. Lee, A new variable step-size NLMS algorithm and its performance analysis, IEEE Trans. Signal Process. 60 (4) (2012) 2055–2060.

[13] J.R. Mohammed, M.S. Shafi, An efficient adaptive noise cancellation scheme using ALE and NLMS filters, Int. J. Electr. Comput. Eng. 2 (3) (2012) 325–332.

[14] R. Martin, Noise power spectral density estimation based on optimal smoothing and minimum statistics, IEEE Trans. Speech Audio Process. 9 (5) (2001) 504–512.

[15] I. Almajai, B. Milner, Visually derived Wiener filters for speech enhancement, IEEE Trans. Audio Speech Lang. Process. 19 (6) (2011) 1642–1651.

[16] T. Aboulnasr, K. Mayyas, A robust variable step-size LMS-type algorithm: analysis and simulations, IEEE Trans. Signal Process. 45 (3) (1997) 631–639.

[17] R. Serizel, M. Moonen, J. Wouters, S.H. Jensen, A zone-of-quiet based approach to integrated active noise control and noise reduction for speech enhancement in hearing aids, IEEE Trans. Audio Speech Lang. Process. 20 (6) (2012) 1685–1697.

[18] S. Haykin, Adaptive Filter Theory, Pearson Education India, 2007.

[19] B.K. Mohanty, P.K. Meher, A high performance energy-efficient architecture for FIR adaptive filter based on new distributed arithmetic formulation of block LMS algorithm, IEEE Trans. Signal Process. 61 (4) (2013) 921–932.

[20] B. Huang, Y. Xiao, J. Sun, G. Wei, A variable step-size FxLMS algorithm for narrowband active noise control, IEEE Trans. Audio Speech Lang. Process. 21 (2) (2013) 301–312.

[21] S.C. Douglas, Fast Implementations of the filtered-X LMS and LMS algorithms for multichannel active noise control, IEEE Trans. Speech Audio Process. 7 (4) (1999) 454–465.

[22] S.M. Kuo, D. Morgan, Active Noise Control Systems: Algorithms and DSP Implementations, John Wiley & Sons, 1996.

[23] J. Benesty, H. Rey, L. Rey Vega, S. Tressens, A nonparametric VSS NLMS algorithm, IEEE Signal Process. Lett. 13 (10) (2006) 581–584.

[24] N.J. Bershad, Analysis of the normalized LMS algorithm with Gaussian inputs, IEEE Trans. Acoust. Speech Signal Process. 34 (4) (1986) 793–806.

CHAPTER 4

Unsupervised single-channel speech enhancement based on phase aware time-frequency mask estimation

Nasir Saleem[a,b] **and Muhammad Irfan Khattak**[b]
[a]Department of Electrical Engineering, Faculty of Engineering & Technology, Gomal University, Dera Ismail Khan, Pakistan
[b]Department of Electrical Engineering, University of Engineering & Technology, Peshawar, Pakistan

4.1 Motivation

There are many forms of human communication, for example, nonverbal and text-based communication. Speech, however, is the most effective and efficient form of human communication. Through speech, we are able to convey instructions, emotions, and so on. The usefulness of speech has led to a variety of speech processing applications. However, successful use of these applications is considerably aggravated in the presence of background noise. The noise signals overlap and mask the useful speech signals. To deal with overlapping background noise, a speech enhancement strategy is essential in order to make noisy speech more understandable and pleasant. Speech enhancement converts noisy speech signals to enhanced speech signals with better perceptual quality and intelligibility. The motivation behind this research work is to deal effectively with background noise and produce high-quality and intelligible enhanced speech.

4.2 Introduction

Speech enhancement improves the performance of digital communications, speech preprocessing for hearing aids, and speech recognition, and it describes an algorithm to enhance perceived speech quality, reduce hearing fatigue, and improve speech intelligibility. The usefulness of speech has led to a variety of speech processing applications such as automatic speech recognition (ASR), hearing aids, and human–machine interaction [1–4].

Applied Speech Processing
https://doi.org/10.1016/B978-01-2-823898-1.00006-0

© 2021 Elsevier Inc.
All rights reserved.

75

However, successful use of these applications is considerably aggravated in the presence of background noise. The noise signals overlap and mask the useful speech signals. To deal with overlapping background noise, a speech enhancement strategy is essential in order to make noisy speech more understandable and pleasant. As such, countless speech enhancement methods have been proposed [5–14]. Large improvements in terms of intelligibility and quality have been obtained by employing a signal-to-noise ratio (SNR)-based binary time-frequency mask. Such binary masks have been formulated to retain the time-frequency units when the speech signals are greater than a predefined threshold (0 dB) (SNR > 0 dB) and discard time-frequency units when speech signals are less than the threshold (SNR > 0 dB). Li and Wang [15] provide an extensive literature on time-frequency masking. Approaches using binary time-frequency masks for speech enhancement have demonstrated significant performance gains in the perceived quality and intelligibility of speech at enormously adverse SNR [16–19]. IBM has been suggested as computational goal of CASA [20]. The above-mentioned studies have encouraged the researchers to develop the novel binary time-frequency masks.

4.3 Literature review

Kang et al. [21] proposed an online source extraction system for speech enhancement based on extracted target signals with source localization constraints. The proposed method enhanced target speech by exploiting the beamforming. An improved time-frequency mask-based BSS approach is proposed to separate speech from multiple noise sources. Wang et al. [6] estimated a time-frequency mask based on the spectral dependency into a speech cochleagram for speech enhancement. To attain spectral dependency, the data field concept has been established to model the time-frequency correlation of the cochleagram, and spectral knowledge has been used to estimate the time-frequency mask. To acquire preliminary time-frequency values of noise and speech signals, a preprocessed module is employed, and a data field model has been employed to acquire the forms of speech and noise potentials. The masking values are achieved by using the potentials of speech and noise for reinstating the clean target speech signal.

Wang [22] developed a novel method to separate target speech from mixtures using time-frequency masking. The study established the time-frequency masking concept and reviewed time-frequency masking algorithms used to separate target speech from both monaural and binaural mixtures. The review emphasized methods capable for hearing-aid design. Another study measured the potential benefits of time-frequency masking methods for the hearing impaired in light of the processing constraints of hearing aids. The research surveyed recent studies that evaluated the perceptual effects of time-frequency masking methods, mostly their success in improving human speech recognition in noisy environments. Bentsen et al. [23] studied the roles and contributions of time-frequency masking by measuring speech intelligibility in normal hearing listeners. A large improvement in speech intelligibility (25.4%) was found when going from a sub-band-based architecture to a deep neural, network-based architecture. A 13.9% improvement was achieved by changing the learning objective from ideal binary mask to ideal ratio mask. Spectral mask estimation-based speech enhancement that uses local binary patterns to estimate an ideal neighborhood mask has been proposed [24]. This method indicated particular time-frequency units of noisy speech as noise-dominated units. A new time-frequency masking method has been proposed in [25] based on sparse approximation and employed a noncausal noise estimator to estimate a time-frequency mask.

Previously, spectral phase estimation was considered trivial for speech enhancement [26], as some believed that phase information is inconsequential during speech perception. Consequently, conventional speech enhancement approaches simply exploited the spectral amplitude and used phase of noisy speech during reconstructing the enhanced speech signal. Recent studies have showed encouraging impacts of phase estimation on the performance of speech enhancement systems [27–31]. A phase-aware complex nonnegative matrix factorization (CMF) method has been proposed [32] to separate overlying parts in the mixture of harmonic music sources. To achieve better performance, it is important to consider phase information during modifications of the spectral amplitudes and the reconstruction of speech signals.

In this study, we incorporated the phase information during reconstruction of speech signals for quality and intelligibility improvements of time-frequency masking-based unsupervised speech enhancement when estimated mask is applied to mixed signals. The main contributions of this study are as follows:

1. A novel nonideal binary mask has been estimated to enhance the noisy speech in order to improve the speech intelligibility and perceptual quality. The estimated mask is nonideal; hence, it can be implemented practically in single-channel speech enhancement.

2. Unlike conventional speech enhancement methods, we have estimated the spectral phase and replaced the noisy phase with the estimated phase, which significantly improved the perceptual speech quality.

The remainder of this chapter is organized as follows. Section 4.4 provides problem definition and notations. Section 4.5 discusses the mechanism of estimating the time-frequency mask, and Section 4.6 discusses phase estimation. Section 4.7 describes the experimental setting and presents the obtained results, and Section 4.8 concludes.

4.4 Problem definition and notations

Lest us assume that noise contaminated speech $y(t)$ is denoted by the summation of underlying speech $s(t)$ and the noise signal $d(t)$, expressed as:

$$y(t) = s(t) + d(t) \tag{4.1}$$

After computing short-time Fourier transform (STFT), we acquire complex spectrums of noisy speech, underlying speech, and noise signals as:

$$Y(\omega, l) = S(\omega, l) + D(\omega, l) \tag{4.2}$$

where ω indicates the frequency units and l indicates the time frame. The complex STFTs contain spectral amplitudes and phases as: $Y(\omega,l) = |Y(\omega,l)| e^{j\phi(\omega,l)}$ and phase $\phi_Y(\omega,l) = \angle Y(\omega,l)$, $S(\omega,l) = |S(\omega,l)| e^{j\phi(\omega,l)}$ and phase $\phi_S(\omega,l) = \angle S(\omega,l)$ whereas $D(\omega,l) = |D(\omega,l)| e^{j\phi(\omega,l)}$ and phase $\phi_D(\omega,l) = \angle D(\omega,l)$, respectively. A speech enhancement estimates an underlying signal $\hat{s}(t)$ by reducing the noise signals. The enhanced speech is restructured by applying an inverse STFT to the estimated signal.

$$\hat{s}_{Enh}(t) = iSTFT\left(|\hat{S}(\omega, l)| e^{j\hat{\phi}(\omega, l)}\right) \tag{4.3}$$

where $\hat{S}(\omega,l)$ and $\hat{\phi}(\omega, l)$ indicate the estimated spectral amplitude and phase. Usually, phase of the noisy speech is used during reconstruction of the enhanced speech. However, it brings artifacts and degrades the quality of enhanced speech [31]. The enhanced spectral amplitude is obtained by using the magnitude mask $M(\omega,l)$ to the spectral amplitude of the noisy speech $|Y(\omega,l)|$ as:

$$|\hat{S}(\omega, l)| = M(\omega, l)|Y(\omega, l)| \tag{4.4}$$

$M(\omega,l)$ denotes the estimated mask. Section 4.5 discusses estimation of the mask.

4.5 Time-frequency mask estimation

Time-frequency masking is a successful method to enhance noisy speech. The ideal binary masking (IdBM) signifies two-dimension binary matrices to mark time-frequency units of mixtures as speech-dominant and noise-dominant [20]. The time-frequency unit with binary 1 is considered to be speech-dominant and contained, whereas the noise-dominant time-frequency unit is eliminated. The proposed method is depicted in Fig. 4.1 where features $f_m(\omega, l)$ are extracted in all frequency units ω. The extracted features discover the peaks and valleys in each ω unit. The features are extracted by calculating the ratio and absolute value of signal variance. The features are computed as:

$$f_m(\omega, l) = 10 \log_{10}\left(\frac{\sigma_{S'}^2(\omega, l)}{\sigma_{|S|}^2(\omega, l)}\right) \tag{4.5}$$

where $\sigma_{S'}^2(\omega, l) = (S(\omega, l))^\alpha$ and $|S(\omega, l)|$ denote the absolute values of L dimension noisy-speech vectors in l frames. The extracted features are smoothed across the time with three-point median filter. To decide noise-free and noise-dominant features in time-frequency units, they are passed through an unsupervised and nonparametric adaptive threshold [33]. The histogram-based threshold estimation is vectors composed of the preceding (L_p) and prospect speech frames (L_f):

$$f_{Hist}(\omega, k) = \left\{ f_N\left(\omega, l - L_p\right), ..., f_N\left(\omega, l + L_f\right)\right\} \tag{4.6}$$

The optimized threshold Tr^* is achieved by using Tr used to maximize $\sigma_Z^2(Tr)$ as:

$$\sigma_Z^2(Tr^*) = \text{Max}\left(\sigma_Z^2(Tr)\right) \tag{4.7}$$

Tr shows summation of the histogram levels for input feature vectors $F_{Hist}(\omega, l)$, and $\sigma_Z^2(Tr)$ is between-class variance with distinct intensity levels Tr given as:

$$\sigma_Z^2(Tr) = \left[\frac{\left[\psi_G \cdot \kappa_s(Tr) - \lambda(Tr)\right]^2}{\kappa_s(Tr)(1 - \kappa_s(Tr))}\right] \tag{4.8}$$

where ψ_G, $\lambda(Tr)$ and $\kappa_s(Tr)$ indicate global mean intensities, cumulative means, and the cumulative sums of the signals, respectively, given as:

$$\psi_G = \sum_{j=1}^{Tr} j \cdot H_j \tag{4.9}$$

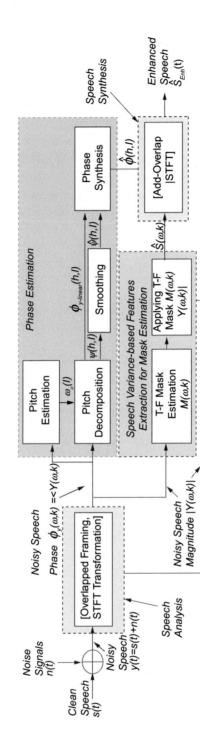

Fig. 4.1 Block diagram of proposed speech enhancement.

$$\lambda(Tr) = \sum_{j=1}^{Tr} j \cdot H_j \tag{4.10}$$

$$\kappa_S(Tr) = \sum_{j=1}^{Tr} H_j \tag{4.11}$$

H_j indicates normalized histograms of the feature vectors. When values in a particular time-frequency unit are greater than the adaptive threshold, the units are considered speech-dominant, whereas when the feature values in time-frequency units are less than the adaptive threshold, the units are considered noise-dominant. The speech-dominant time-frequency units are held, whereas the noise-dominant units are removed. The resulting mask $M(\omega,l)$ is given as:

$$M(k, l_k) = \begin{cases} 1 & \text{if, } f_m(k, l_k) > \max\left(Tr^*(k, l_k), Tr_0\right) \\ 0 & \text{otherwise} \end{cases} \tag{4.12}$$

where Tr_0 is the minimum threshold level. Following the selection of time-frequency units, inverse STFT is used to reconstruct the enhanced speech using estimated phase of underlying clean speech. Section 4.6 describes the process of phase estimation.

4.6 Phase estimation

A phase estimation method proposed in Ref. [27] is combined with the estimated mask presented in Section 4.3. The phase estimator decomposes the noisy phase spectrum into minimum phase, linear phase (LP), and dispersion phase. When LP is taken off the instantaneous phase, the noise components can be reduced by employing a temporal smoothing filter (TSF) over the remaining unwrapped phase. This minimizes estimation error variance of the noisy phase. The input noisy speech $y(t,l)$ is divided into frames $y_w(t,l)$. Each frame $y_w(t,l)$ in voiced parts are replicated as summation of the harmonics containing the spectral amplitudes and instantaneous phases, given as:

$$y_w(t, l) \approx \sum_{h=1}^{H} A(h, l) \cos\left(h\omega_0(l) + \phi_y(h, l)\right) \tag{4.13}$$

where $A(t,l)$ is the spectral amplitude and $h \in [1, H_l]$ denotes the harmonic index with H_l at frame l while $\omega_0(l) = 2\pi f_o(l)/f_s$, where f_o is the fundamental frequency and f_s is sampling frequency. The harmonic phase $\phi_y(h,l)$ can be decomposed as [34]:

$$\phi_y(h, l) = h \sum_{l'=0}^{l} \omega_o(l')(t(l') - t(l'-1)) + \angle V(h, l) + \phi_n(h, l) \quad (4.14)$$

In Eq. (4.14), the first term is LP given as:

$$\phi_{y-linear}(h, l) = h \sum_{l'=0}^{l} \omega_o(l')(t(l') - t(l'-1)) \quad (4.15)$$

The second term is minimum phase denoted by $V(h,l)$. Normally, the instantaneous phase is wrapped by the linear phase over time. The last part called the dispersion phase detains the stochastic phase in the excitation signals. The decomposition in Eq. (4.14) decouples the instantaneous phase into the linear phase (LP) and unwrapped phase (UwP), respectively. The LP is ascertained and does not contain vital perceptual contents except f_o [35]. Once speech is degraded by background noise, the UwP components also become noisy. Consequently, estimated LP is subtracted from the instantaneous phase. TSF is applied on the residual UwP. The UwP is attained by eliminating the LP estimate as:

$$\psi(h, l) = \phi_{y-linear}(h, l) - \phi_y(h, l) \quad (4.16)$$

Afterward, the UwP is smoothed along time by using following equation:

$$\hat{\psi}(h, l) = \angle \sum_{l'=l-N/2}^{l+N/2} e^{j\psi(h, l')} \quad (4.17)$$

where N shows frame numbers. The estimated UwP is used for reconstruction of enhanced speech as:

$$\hat{\phi}(h, l) = \hat{\psi}(h, l) + \phi_{y-linear}(h, l) \quad (4.18)$$

4.7 Experimental settings

We rigorously examined the proposed unsupervised using various objective measurements. In experiments, noisy speech utterances are produced by mixing different background noises with clean speech utterance at a range of global SNRs. We randomly selected 200 clean speech utterances from the Institute of Electrical and Electronics Engineers (IEEE) database [36]. We selected the background noise sources from the Aurora database [37], which includes airport noise, babble noise, car noise, exhibition hall noise, street noise, and white noise. Fig. 4.2 depicts the spectrograms of

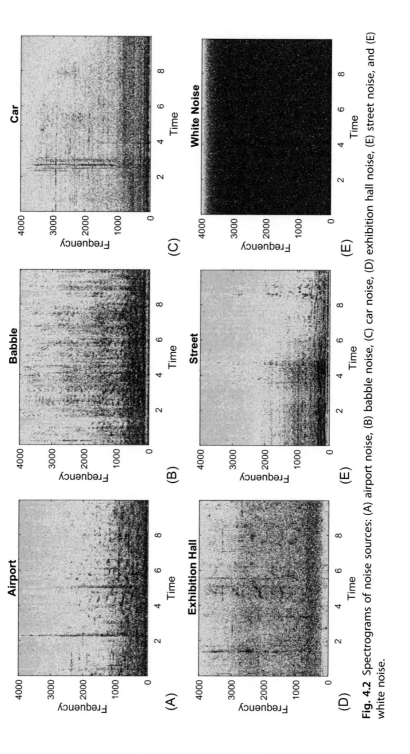

Fig. 4.2 Spectrograms of noise sources: (A) airport noise, (B) babble noise, (C) car noise, (D) exhibition hall noise, (E) street noise, and (E) white noise.

noise sources. The noisy utterances are generated at four SNRs ranging from 0 to 10 dB at 5 dB spacing utilizing the ITU-T P.51 standard and re-sampled to a frequency of 8 kHz. Three conventional speech enhancement methods including CASA-based IBM [17], DNN-based IBM [38], and Weiner filtering (WF) [39] are used for comparison purpose. The proposed mask with blind phase is denoted by pIBM while the mask with estimated phase is denoted by pIBM + PE. We used four objective metrics to access performance. The ITU-T Recommendation P.862, perceptual evaluation of speech quality (PESQ) [40] is used to measure the quality of enhanced speech. PESQ values are calculated as linear summations of symmetrical and asymmetrical disturbances, d_{sym} and d_{Asym} as:

$$PESQ = \lambda_o + \lambda_1 \cdot d_{sym} + \lambda_2 \cdot d_{Asym} \tag{4.19}$$

where $\lambda_o = 4.5$, $\lambda_1 = -0.1$ and $\lambda_2 = -0.039$. PESQ value pursues MOS; a large value indicates a better performance. The SNR measure is an accepted method to assess the enhanced speech. However, the standard SNR metric shows poor correlation with speech quality because this metric computes the average ratio of the entire speech signal. Speech signals exhibit high nonstationary nature; therefore, swift rise-fall in the speech signals causes erroneous SNR calculations. Few portions may compose useful speech parts, whereas the noise parts remain inaudible and vice versa. Therefore, entire signal averages may eliminate critical speech parts. To circumvent this problem, SNRs are calculated in small segments and averaged. This metric is used to assess the amount of noise in the enhanced speech and is referred to as segmental SNR (SSNR).

$$SSNR = \frac{10}{W} \sum_{w=0}^{W-1} \log_{10} \frac{\sum_{t=K_m}^{t=K_m + K - 1} x^2(t)}{\sum_{t=K_m}^{t=K_m + K - 1} [x(t) - \hat{x}(t)]^2} \tag{4.20}$$

where K and W indicate length and number of frames. Short-time objective intelligibility (STOI) [41], a correlation-based method used to access intelligibility of enhanced speech, has shown correlation with the subjective evaluation of intelligibility. Here, the intelligibility values are calculated using the mapping function employed in Ref. [41].

$$f(\alpha) = \frac{100}{1 + \exp(v\alpha + \eta)} \tag{4.21}$$

where $v = -17.4906$ and $\eta = 9.6921$. To access speech distortion, ITU-T Recommendation P.835 (C_{SIG}) [43] is employed, which is the combination of several objective measures, given as:

$$C_{\text{SIG}} = 2.164 - 0.02 \cdot \text{IS} + 0.832 \cdot \text{PESQ} - 0.494 \cdot \text{CEP} + 0.352 \cdot \text{LLR}$$

$$(4.22)$$

where IS denotes the Itakura Saito distance measure [42], CEP denotes the Cepstrum distance measure [42], and LLR denotes the log–likelihood ratio [42], respectively. The experimental conditions are simulated during experiments and are given in Table 4.1.

4.8 Results and discussion

To confirm the effectiveness of the proposed unsupervised speech enhancement method in various noisy backgrounds, we judged the results with three contending methods. Table 4.2 shows the PESQ values obtained with pIBM, pIBM + PE, and competing methods. The best PESQ values were obtained with pIBM + PE, which are consistently greater than competing methods in all conditions. Excellent PESQ scores were achieved with pIBM + PE at low SNRs (0 dB). In terms of PESQ, pIBM-based speech enhancement performed better than CASA-IBM and DNN-IBM in all noisy environments. For example, the calculated PESQ scores with airport

Table 4.1 Experimental setup.

1	Clean speech utterances	IEEE database [36]
2	Noise sources	Aurora database [37]
3	Sampling frequency	8000 samples/s
4	Frame duration	30 ms
5	Windowing function	Hann window
6	Window overlap	50% Overlap
7	FFT length	512 Samples
8	Competing methods	CASA-IBM [17]
		DNN-IBM [38]
		Weiner filtering [39]
9	Input SNRs	0 dB, 5 dB, 10 dB
10	Objective evaluation measures	PESQ [40]
		SSNR [42]
		STOI [41]
		C_{SIG} [43]
11	Noisy stimuli generation	ITU-T P.51 Recommendation

Table 4.2 Performance evaluation in terms of PESQ.

Noise	Methods	0 dB	5 dB	10 dB
Airport noise	Noisy	1.81	1.98	2.51
	WF	1.93	2.28	2.78
	CASA-IBM	2.41	2.74	2.94
	DNN-IBM	2.23	2.49	2.87
	pIBM	2.57	2.81	3.11
	pIBM + PE	2.63	2.95	3.19
Car noise	Noisy	1.92	2.08	2.62
	WF	2.01	2.39	2.89
	CASA-IBM	2.48	2.81	2.98
	DNN-IBM	2.37	2.55	2.91
	pIBM	2.61	2.87	3.09
	pIBM + PE	2.69	2.98	3.22
Street noise	Noisy	1.70	1.85	2.23
	WF	1.92	2.17	2.48
	CASA-IBM	2.39	2.71	2.91
	DNN-IBM	2.26	2.51	2.85
	pIBM	2.54	2.79	3.10
	pIBM + PE	2.66	2.98	3.17
Babble noise	Noisy	1.80	2.06	2.38
	WF	1.98	2.29	2.78
	CASA-IBM	2.45	2.71	2.91
	DNN-IBM	2.15	2.57	2.75
	pIBM	2.58	2.89	3.05
	pIBM + PE	2.68	2.97	3.18
Exhibition hall	Noisy	1.72	2.29	2.43
	WF	1.91	2.41	2.58
	CASA-IBM	2.32	2.67	2.81
	DNN-IBM	2.19	2.53	2.81
	pIBM	2.48	2.78	3.02
	pIBM + PE	2.57	2.89	3.14
White noise	Noisy	1.61	1.81	2.17
	WF	1.98	2.32	2.67
	CASA-IBM	2.52	2.83	3.01
	DNN-IBM	2.42	2.57	2.98
	pIBM	2.64	2.91	3.23
	pIBM + PE	2.71	3.05	3.31

noise improved from 2.23 with DNN-IBM to 2.57 with pIBM (Δ PESQ $= 0.34$) and 2.63 with pIBM + PE (Δ PESQ $= 0.40$) at 0 dB. Similarly, the calculated PESQ scores with white noise improved from 2.52 with CASA–IBM to 2.64 with pIBM (Δ PESQ $= 0.12$) and 2.71 with pIBM + PE (Δ PESQ $= 0.19$) at 0 dB. In the same way, the calculated PESQ scores with babble noise improved from 1.81 with un-processed speech to 2.27 with pIBM (Δ PESQ $= 0.76$) and 2.68 with pIBM + PE (Δ PESQ $= 0.87$) at 0 dB. In addition, the calculated PESQ scores with car noise improved from 2.01 with WF to 2.61 with pIBM (Δ PESQ $= 0.60$) and 2.69 with pIBM + PE (Δ PESQ $= 0.68$) at 0 dB. Figs. 4.2B and 4.3A show delta PESQ improvements (Δ PESQ) for pIBM and pIBM + PE in all noise sources while Fig. 4.4 shows average Δ PESQ for pIBM + PE against competing methods. From Δ PESQ results, it is clear

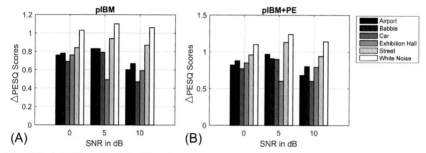

Fig. 4.3 Δ PESQ analysis in all noise sources.

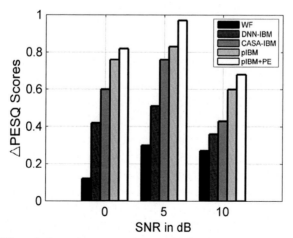

Fig. 4.4 Δ PESQ analysis against competing methods.

that pIBM and pIBM + PE outperformed the competing methods at all SNRs and reflected high-quality speech. Analysis of variance (ANOVA) statistical analysis was carried out in order to confirm the significance difference in PESQ values. Major improvements are noted with pIBM and pIBM + PE ($P < 0.0001$) at 0 dB SNR in all noisy environments. At high SNRs (5 and 10 dB), notable improvements in PESQ values were observed excluding exhibition hall ($P < 0.031$). In comparison to competing methods, pIBM and pIBM + PE performed very well in all noisy environments except 10 dB exhibition hall ($P < 0.05$). According to Tukey HSD post hoc test, PESQ values achieved with pIBM and pIBM + PE are confirmed significant at all input SNR levels.

In order to observe suppression of noise in the reconstructed enhanced speech, we used SSNR. Table 4.3 illustrates the SSNR results achieved with pIBM, pIBM + PE, and competing methods, respectively. High SSNR scores are achieved with pIBM and pIBM + PE. For pIBM, the calculated SSNR scores improved from 2.43 with CASA-IBM to 2.67 ($\Delta SSNR = 0.24$) at 0 dB airport noise and calculated SSNR scores improved from 2.51 with DNN-IBM to 2.64 ($\Delta SSNR = 0.13$) at 0 dB babble noise, respectively. For pIBM + PE, the computed SSNR scores improved from 1.15, 2.92, 3.12, and 3.21 with noisy speech, WF, CASA-IBM, and DNN-IBM to 3.46 at 0 dB white noise: $\Delta SSNR = 2.31$, $\Delta SSNR = 0.54$, $\Delta SSNR = 0.34$, and $\Delta SSNR = 0.25$, respectively. The highest progress in SSNR ($\Delta_{HIGH}SSNR = 2.31$) is noted at 0 dB white noise, whereas the lowest progress ($\Delta_{LOW}SSNR = 2.31$) is observed at 10 dB airport noise. The SSNR improvements ($\Delta SSNR$) achieved with pIBM and pIBM + PE are depicted in Figs. 4.4B and 4.5A. Fig. 4.6 shows average $\Delta SSNR$ for pIBM + PE against competing methods. Again, ANOVA-based statistical analysis showed significant improvements with pIBM and pIBM + PE ($P < 0.0001$) at 0 dB SNR. At 5 and 10 dB, considerable improvements in SSNR scores are observed except street noise ($P < 0.042$). Tukey's HSD post hoc test confirmed that SSNR scores achieved with pIBM and pIBM + PE are significant ($P < 0.05$) at all input SNRs except 10 dB street noise. Speech enhancement is mainly used to reduce noise from contaminated speech in order to improve speech quality for communication and playback. However, in speech recognition, speech intelligibility is a more important aspect. Speech intelligibility indicates understanding of individual spoken items. Table 4.4 shows speech intelligibility scores achieved using STOI where pIBM- and pIBM + PE-based speech enhancement beat other methods and led to overall intelligibility

Table 4.3 Performance evaluations in terms of SSNR.

Noise	Methods	0 dB	5 dB	10 dB	Noise	Methods	0 dB	5 dB	10 dB
Airport noise	Noisy	1.23	2.66	6.08	Babble noise	Noisy	1.17	2.81	5.16
	WF	2.17	3.58	6.71		WF	2.01	3.91	6.05
	CASA-IBM	2.43	4.04	6.87		CASA-IBM	2.42	4.01	6.67
	DNN-IBM	2.55	4.39	6.99		DNN-IBM	2.51	4.17	6.71
	pIBM	2.67	4.51	7.07		pIBM	2.64	4.29	6.86
	pIBM + PE	2.83	4.65	7.19		pIBM + PE	2.87	4.42	6.97
Car noise	Noisy	1.21	2.62	4.63	Exhibition hall	Noisy	1.13	3.60	4.98
	WF	2.42	4.01	6.31		WF	2.10	4.05	6.19
	CASA-IBM	2.48	4.23	6.58		CASA-IBM	2.32	4.34	6.61
	DNN-IBM	2.37	4.39	6.71		DNN-IBM	2.19	4.48	6.82
	pIBM	2.61	4.51	6.89		pIBM	2.48	4.57	6.93
	pIBM + PE	2.86	4.64	7.01		pIBM + PE	3.11	4.72	7.09
Street noise	Noisy	1.29	2.73	5.38	White noise	Noisy	1.15	2.81	4.87
	WF	2.18	4.07	6.45		WF	2.92	4.51	6.53
	CASA-IBM	2.39	4.33	6.61		CASA-IBM	3.12	4.71	6.71
	DNN-IBM	2.26	4.47	6.75		DNN-IBM	3.21	4.77	6.88
	pIBM	2.54	4.53	6.87		pIBM	3.43	4.86	6.97
	pIBM + PE	2.46	4.69	6.98		pIBM + PE	3.46	5.01	7.13

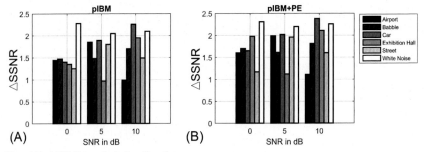

Fig. 4.5 △SSNR analysis in all noise sources.

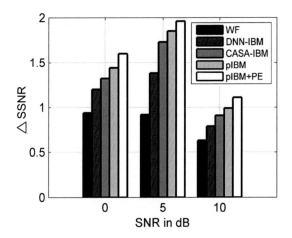

Fig. 4.6 △SSNR analysis against competing methods.

STOI > 80% for SNR ≥ 0 dB. The pIBM + PE showed preeminent overall intelligibility of 95%. Note from Table 4.4 that pIBM + PE obtained the finest average scores for all noisy backgrounds. For illustration, the average calculated STOI scores with airport noise improved from 72% with CASA-IBM and 76% with DNN-IBM to 78% with pIBM and 81% with pIBM + PE at 0 dB. Likewise, the average values with white noise increased from 71% with WF to 79% with pIBM and 81% with pIBM + PE at 0 dB. Overall, the average STOI values in all noisy backgrounds for pIBM are 79%, 87.5%, and 91.33% at 0, 5, and 10 dB, respectively. Similarly, the overall average STOI scores in all noises for pIBM + PE are 82%, 91%, and 94% at 0, 5, and 10 dB, respectively. Fig. 4.7A and B shows STOI improvements (△STOI) for pIBM and pIBM + PE in all noisy conditions.

Fig. 4.8 shows the average △STOI for pIBM + PE against competing methods. According to △STOI, it is clear that pIBM and pIBM + PE

Table 4.4 Performance evaluations in terms of STOI.

Noise	Methods	0 dB	5 dB	10 dB	Noise	Methods	0 dB	5 dB	10 dB
Airport noise	Noisy	0.62	0.73	0.81	Babble noise	Noisy	0.68	0.76	0.83
	WF	0.67	0.78	0.86		WF	0.71	0.81	0.88
	CASA-IBM	0.72	0.81	0.85		CASA-IBM	0.75	0.84	0.87
	DNN-IBM	0.75	0.85	0.88		DNN-IBM	0.78	0.87	0.89
	pIBM	0.78	0.88	0.90		pIBM	0.80	0.91	0.91
	pIBM + PE	0.81	0.91	0.93		pIBM + PE	0.85	0.95	0.94
Car noise	Noisy	068.	0.78	0.82	Exhibition hall	Noisy	0.64	0.73	0.84
	WF	0.71	0.83	0.88		WF	0.67	0.78	0.88
	CASA-IBM	0.74	0.84	0.88		CASA-IBM	0.72	0.80	0.89
	DNN-IBM	0.76	0.86	0.89		DNN-IBM	0.75	0.83	0.91
	pIBM	0.79	0.89	0.91		pIBM	0.78	0.85	0.93
	pIBM + PE	0.82	0.92	0.94		pIBM + PE	0.80	0.88	0.95
Street noise	Noisy	0.67	0.75	0.83	White noise	Noisy	0.69	0.78	0.84
	WF	0.71	0.79	0.87		WF	0.71	0.82	0.88
	CASA-IBM	0.75	0.81	0.88		CASA-IBM	0.74	0.84	0.89
	DNN-IBM	0.78	0.83	0.89		DNN-IBM	0.77	0.87	0.91
	pIBM	0.80	0.85	0.90		pIBM	0.79	0.90	0.93
	pIBM + PE	0.82	0.88	0.93		pIBM + PE	0.81	0.92	0.95

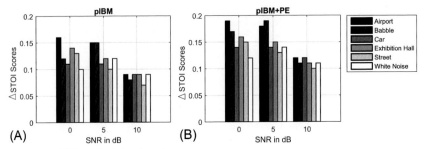

Fig. 4.7 △STOI analysis in all noise sources.

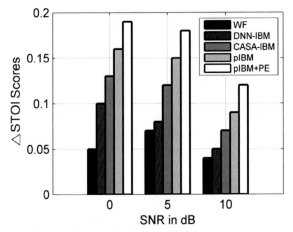

Fig. 4.8 △STOI analysis against competing methods.

outperformed other methods and revealed the greatest intelligibility. Again we conducted ANOVA-based statistical analysis, which showed that the improvements observed with pIBM and pIBM + PE ($P < 0.0001$) are significant. Notable speech intelligibility improvements are highlighted at all input SNRs except exhibition hall ($P < 0.027$) and street ($P < 0.0261$). Likewise, in contrast to the competing methods, pIBM and pIBM + PE performed the best apart from 10 dB exhibition hall ($P < 0.05$) and 10 dB street ($P < 0.05$). The Tukey's HSD post hoc test confirmed that the STOI values obtained with pIBM and pIBM + PE are statistically significant ($P < 0.05$).

In order to inspect the distortion in the reconstructed speech, the composite measure (C_{SIG}) is considered. Table 4.5 shows the C_{SIG} values obtained with pIBM, pIBM + PE, and competing methods. High C_{SIG} values are obtained with pIBM and pIBM + PE. For pIBM + PE, the

Table 4.5 C_{SIG} scores in various noisy environments.

Noise	Methods	0 dB	5 dB	10 dB	Noise	Methods	0 dB	5 dB	10 dB
Airport noise	Noisy	1.91	2.34	2.88	Babble noise	Noisy	1.47	2.27	2.38
	WF	2.32	2.66	3.22		WF	2.10	2.61	3.08
	CASA–IBM	2.52	2.93	3.54		CASA–IBM	2.35	2.95	3.29
	DNN–IBM	2.61	2.97	3.66		DNN–IBM	2.55	2.94	3.71
	pIBM	2.76	3.09	3.87		pIBM	2.88	3.16	3.74
	pIBM + PE	2.84	3.15	3.93		pIBM + PE	2.91	3.27	3.98
Car noise	Noisy	1.89	2.44	2.88	Exhibition hall	Noisy	1.72	2.29	2.82
	WF	2.41	2.73	3.22		WF	2.29	2.57	3.14
	CASA–IBM	2.61	3.03	3.63		CASA–IBM	2.43	2.87	3.48
	DNN–IBM	2.72	3.17	3.72		DNN–IBM	2.51	2.89	3.63
	pIBM	2.88	3.29	3.97		pIBM	2.64	3.01	3.85
	pIBM + PE	2.95	3.21	4.01		pIBM + PE	2.79	3.11	3.94
Street noise	Noisy	1.83	2.30	2.79	White noise	Noisy	1.98	2.41	2.91
	WF	2.29	2.56	3.01		WF	2.42	2.72	3.31
	CASA–IBM	2.48	2.82	3.49		CASA–IBM	2.61	2.99	3.62
	DNN–IBM	2.57	2.88	3.61		DNN–IBM	2.72	3.17	3.77
	pIBM	2.69	2.93	3.74		pIBM	2.81	3.29	3.92
	pIBM + PE	2.78	3.09	3.88		pIBM + PE	2.96	3.31	4.03

calculated average C_{SIG} values improved from 2.86 with CASA-IBM to 3.34 ($\Delta C_{SIG} = 0.48$) during babble noise and computed average C_{SIG} values improved from 3.22 with DNN-IBM to 3.43 ($\Delta C_{SIG} = 0.21$) during white noise. Similarly, with pIBM + PE, the computed average C_{SIG} values improved from 2.43 with WF to 2.96 ($\Delta C_{SIG} = 0.50$) during car noise. ANOVA statistical analysis showed significant improvements with pIBM and pIBM + PE ($P < 0.0001$) at 0, 5, and 10 dB SNR. Tukey's HSD post hoc test confirmed that C_{SIG} values achieved with pIBM and pIBM + PE are significant. The fluctuations in estimating the noise PSD can introduce unavoidable fluctuations and such details may produce musical noise [43]. In contrast, fewer fluctuations during noise PSD estimation, as illustrated in Fig. 4.9, may lead to a small amount of variations in the enhanced speech. As such, it is possible to decrease the quantity of the musical noise and consequently improve the overall quality of the speech.

For further understanding about residual background noise and distortion in the enhanced speech, we examined the T-F distribution of the enhanced speech and assessed the enhancement methods for residual background noise and speech distortion by investigating the spectrograms of the speech utterances. Fig. 4.10 shows the spectrograms of various

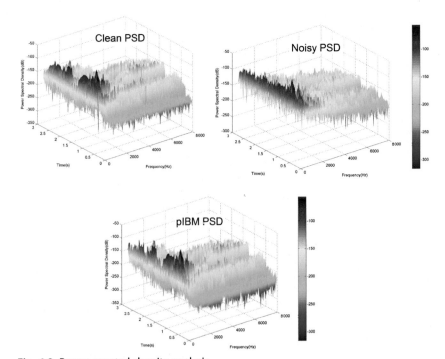

Fig. 4.9 Power spectral density analysis.

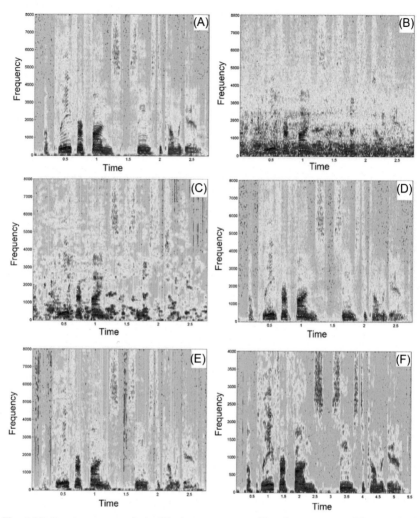

Fig. 4.10 Spectrogram analysis: (A) clean utterance, (B) noisy utterance, (C) processed by Wiener filtering, (D) processed by CASA-IBM, (E) processed by DNN-IBM, and (F) processed by pIBM + PE.

methods. An example speech utterance contaminated by street noise at 0 dB is investigated. As shown in Fig. 4.10C–F, the harmonic structures of the vowels are retained in enhanced speech. Consequently, the processing methods did not experience overattenuation. Moreover, the spectrograms showed an excellent shape in speech activity. In the speech-pause parts, our method is effective in reducing the background noises, marked by boxes. Weak harmonics in high bands are maintained by the proposed method and thus showed a better speech quality. In contrast, a residual background noise is depicted in the spectrograms as shown in Fig. 4.10C–E. The residual background noise is minimized by our method as shown in Fig. 4.10F. The weak energy parts are preserved and capitulated low speech distortion.

4.9 Conclusion

In this chapter, we proposed an unsupervised speech enhancement technique that combines time-frequency masking with phase estimation in order to minimize low-frequency noise from contaminated speech signals. Conventionally, speech enhancement techniques use phase of noisy speech during reconstruction. However, phase information can significantly improve speech quality and intelligibility. We constructed a time-frequency mask and combined it with estimated phase of underlying clean speech. We extracted variance-based features to estimate the time-frequency mask, and weighed the speech features against an adaptive threshold. The features satisfying the condition, $f_m(\omega,k) > \max(Tr^*(\omega,k), Tr_o)$, were held, and the features violating the condition, $f_m(\omega,k) < \max(Tr^*(\omega,k), Tr_o)$, were removed. The estimated magnitude mask was used to obtain a reduced noise speech. Customary objective matrices were employed in order to determine the quality and intelligibility of the reconstructed speech signals. We used ANOVA for statistical analysis of the obtained PESQ, SSNR, STOI, and C_{SIG} values. The pIBM and pIBM + PE achieved greater PESQ and SSNR values. Considerable improvements in terms of PESQ and SegSNR were achieved with pIBM and pIBM + PE and outperformed competing speech enhancement methods. Statistical improvements were obtained with pIBM and pIBM + PE at low SNRs in all noisy environments. Tukey HSD confirmed that the scores achieved with pIBM + PE were statistically significant as compared to DNN-IBM and CASA-IBM. Spectral analysis demonstrated that pIBM and pIBM + PE greatly decreased the residual background noise and maintained the speech parts well. The conclusions of the study are vital since the fallouts suggested that pIBM- and pIBM + PE-based speech enhancement showed immense potential to improve speech quality and intelligibility. The proposed method significantly improved the speech quality and intelligibility of speech in many noisy conditions; however, speech recognition potentials of the proposed method need to be verified in such situations.

References

[1] S. Sen, A. Dutta, N. Dey, Audio Processing and Speech Recognition: Concepts, Techniques and Research Overviews, Springer, 2019.

[2] N. Dey (Ed.), Intelligent Speech Signal Processing, Academic Press, 2019.

[3] S. Sen, A. Dutta, N. Dey, Speech processing and recognition system, in: Audio Processing and Speech Recognition, Springer, Singapore, 2019, pp. 13–43.

[4] N. Dey, A.S. Ashour, W.S. Mohamed, N.G. Nguyen, Acoustic sensors in biomedical applications, in: Acoustic Sensors for Biomedical Applications, Springer, Cham, 2019, pp. 43–47.

[5] F. Bao, W.H. Abdulla, A new time-frequency binary mask estimation method based on convex optimization of speech power, Speech Commun. 97 (2018) 51–65.

[6] X. Wang, F. Bao, C. Bao, IRM estimation based on data field of cochleagram for speech enhancement, Speech Commun. 97 (2018) 19–31.

[7] N. Saleem, M. Irfan, Noise reduction based on soft masks by incorporating SNR uncertainty in frequency domain, Circuits Syst. Signal Process. 37 (2018) 2591–2612.

[8] Y. Zhu, X. Pan, Robust frequency invariant beamforming with low sidelobe for speech enhancement, J. Phys. Conf. Ser. 1 (2018), 012043.

[9] H. Seo, M. Lee, J.H. Chang, Integrated acoustic echo and background noise suppression based on stacked deep neural networks, Appl. Acoust. 133 (2018) 194–201.

[10] L. Fiedler, M. Wöstmann, C. Graversen, A. Brandmeyer, T. Lunner, J. Obleser, Single-channel in-ear-EEG detects the focus of auditory attention to concurrent tone streams and mixed speech, J. Neural Eng. 14 (2017), 036020.

[11] N. Saleem, M. Shafi, E. Mustafa, A. Nawaz, A novel binary mask estimation based on spectral subtraction gain-induced distortions for improved speech intelligibility and quality, Tech. J. 20 (2015) 36. University of Engineering and Technology Taxila.

[12] L. Zao, R. Coelho, P. Flandrin, Speech enhancement with emd and hurst-based mode selection, IEEE/ACM Trans. Audio Speech Lang. Process. 22 (2014) 899–911.

[13] R.M. Nickel, R.F. Astudillo, D. Kolossa, R. Martin, Corpus-based speech enhancement with uncertainty modeling and cepstral smoothing, IEEE Trans. Audio Speech Lang. Process. 21 (2013) 983–997.

[14] N. Saleem, M.I. Khattak, G. Witjaksono, G. Ahmad, Variance based time-frequency mask estimation for unsupervised speech enhancement, Multimed. Tools Appl. 78 (22) (2019) 31867–31891.

[15] Y. Li, D. Wang, On the optimality of ideal binary time–frequency masks, Speech Commun. 51 (3) (2009) 230–239.

[16] D. Wang, U. Kjems, M.S. Pedersen, J.B. Boldt, T. Lunner, Speech intelligibility in background noise with ideal binary time-frequency masking, J. Acoust. Soc. Am. 125 (4) (2009) 2336–2347.

[17] N. Li, P.C. Loizou, Factors influencing intelligibility of ideal binary-masked speech: implications for noise reduction, J. Acoust. Soc. Am. 123 (3) (2008) 1673–1682.

[18] U. Kjems, J.B. Boldt, M.S. Pedersen, T. Lunner, D. Wang, Role of mask pattern in intelligibility of ideal binary-masked noisy speech, J. Acoust. Soc. Am. 126 (3) (2009) 1415–1426.

[19] N. Madhu, A. Spriet, S. Jansen, R. Koning, J. Wouters, The potential for speech intelligibility improvement using the ideal binary mask and the ideal wiener filter in single channel noise reduction systems: application to auditory prostheses, IEEE Trans. Audio Speech Lang. Process. 21 (1) (2012) 63–72.

[20] G.J. Brown, D. Wang, Separation of speech by computational auditory scene analysis, in: Speech Enhancement, Springer, Berlin, Heidelberg, 2005, pp. 371–402.

[21] M.A. Kang, S. Jeong, M. Hahn, On-line speech enhancement by time-frequency masking under prior knowledge of source location, Int. J. Comput. Sci. Eng. 1 (2007) 2.

[22] D. Wang, Time-frequency masking for speech separation and its potential for hearing aid design, Trends Amplif. 12 (4) (2008) 332–353.

[23] T. Bentsen, T. May, A.A. Kressner, T. Dau, The benefit of combining a deep neural network architecture with ideal ratio mask estimation in computational speech segregation to improve speech intelligibility, PLoS One 13 (5) (2018).

[24] C.D. Arcos, M. Vellasco, A. Alcaim, Ideal neighbourhood mask for speech enhancement, Electron. Lett. 54 (5) (2018) 317–318.

[25] A.A. Kressner, D.V. Anderson, C.J. Rozell, Causal binary mask estimation for speech enhancement using sparsity constraints, Proc. Meet. Acoust. 19 (1) (2013) 055037.

[26] P. Vary, Noise suppression by spectral magnitude estimation mechanism and theoretical limits, Signal Process. 8 (1985) 387–400.

[27] J. Kulmer, P. Mowlaee, Phase estimation in single channel speech enhancement using phase decomposition, IEEE Signal Process. Lett. 22 (2015) 598–602.

[28] P. Mowlaee, J. Kulmer, Harmonic phase estimation in single-channel speech enhancement using phase decomposition and SNR information, IEEE/ACM Trans. Audio Speech Lang. Process. 23 (2015) 1521–1532.

[29] P. Mowlaee, J. Kulmer, Phase estimation in single-channel speech enhancement: limits-potential, IEEE Trans. Audio Speech Lang. Process. 23 (2015) 1283–1294.

[30] P. Mowlaee, R. Saeidi, Time-frequency constraints for phase estimation in single-channel speech enhancement, in: 2014 14th International Workshop on Acoustic Signal Enhancement (IWAENC), 2014, pp. 337–341.

[31] P. Mowlaee, R. Saeidi, R. Martin, Phase estimation for signal reconstruction in single-channel source separation, in: Thirteenth Annual Conference of the International Speech Communication Association, 2012.

[32] J. Bronson, P. Depalle, Phase constrained complex NMF: separating overlapping partials in mixtures of harmonic musical sources, in: 2014 IEEE International Conference on Acoustics, Speech and Signal Processing (ICASSP), May, IEEE, 2014, pp. 7475–7479.

[33] N. Otsu, A threshold selection method from gray-level histograms, IEEE Trans. Syst. Man Cybern. 9 (1979) 62–66.

[34] Y. Agiomyrgiannakis, Y. Stylianou, Wrapped Gaussian mixture models for modeling and high-rate quantization of phase data of speech, IEEE Trans. Audio Speech Lang. Process. 17 (2009) 775–786.

[35] G. Degottex, D. Erro, A measure of phase randomness for the harmonic model in speech synthesis, in: Fifteenth Annual Conference of the International Speech Communication Association, 2014.

[36] E.H. Rothauser, IEEE recommended practice for speech quality measurements, IEEE Trans. Audio Electroacoust. 17 (1969) 225–246.

[37] H.G. Hirsch, D. Pearce, The Aurora experimental framework for the performance evaluation of speech recognition systems under noisy conditions, in: ASR2000—Automatic Speech Recognition: Challenges for the New Millenium ISCA Tutorial and Research Workshop (ITRW), 2000.

[38] Y. Wang, A. Narayanan, D. Wang, On training targets for supervised speech separation, IEEE/ACM Trans. Audio Speech Lang. Process. 22 (2014) 1849–1858.

[39] P. Scalart, Speech enhancement based on a priori signal to noise estimation, in: IEEE International Conference on Acoustics, Speech, and Signal Processing, 1996 (ICASSP-96), Conference Proceedings, May, vol. 2, 1996, pp. 629–632.

[40] A.W. Rix, J.G. Beerends, M.P. Hollier, A.P. Hekstra, Perceptual evaluation of speech quality (PESQ)—a new method for speech quality assessment of telephone networks and codecs, in: 2001 IEEE International Conference on Acoustics, Speech, and Signal Processing (ICASSP'01), Proceedings, vol. 2, 2001, pp. 749–752.

[41] C.H. Taal, R.C. Hendriks, R. Heusdens, J. Jensen, An algorithm for intelligibility prediction of time–frequency weighted noisy speech, IEEE Trans. Audio Speech Lang.
Process. 19 (2011) 2125–2136.
[42] P.C. Loizou, Speech Enhancement: Theory and Practice, CRC Press, 2013.
[43] ITU-T Recommendation, Subjective Test Methodology for Evaluating Speech Communication Systems That Include Noise Suppression Algorithm, ITU-T Recommendation, 2003, p. 835.

CHAPTER 5

Harmonic adaptive speech synthesis

Mahdi Khosravy[a], Mohammad Reza Alsharif[b], Linnan Zhang[c], and Faramarz Alsharif[d]

[a]Media Integrated Communication Laboratory, Graduate School of Engineering, Osaka University, Suita, Osaka, Japan
[b]University of the Ryukyus, Ryukyus, Okinawa, Japan
[c]NTT Data Global Service, Tokyo, Japan
[d]Kitami Institute of Technology, Kitami, Hokkaido, Japan

5.1. Introduction

Speech generation means generation of a high-level, model speech audio wave from a low-level sequence of features. Generation of speech according to the available speech features is the main approach to artificially synthesize speech. Synthetic speech is applicable for different practical purposes, the simplest being making a robot talk. In other words, machine intelligence assistance for a tactile machine-human connection. Simply, whenever the machine/robot needs to talk to a human user, synthetic speech is used.

This chapter presents a novel approach to speech synthesis based on an adaptive harmonic filter (AHF) [1, 2]. The proposed speech synthesis approach is inspired by the harmonics structure of real speech, which is composed of the natural oscillations in the vocal tract. AHF synthesizes speech by the features adaptively learned from the corresponding real speech. The technique was invented by Asharif's digital signal processing laboratory's [3] research works on adaptive filtering and its application to echo cancellation and interference estimation, which dates back to early work on TV ghost cancellation [4], and technical researches in Fujitsu laboratories on echo cancellation [5]. These research works on adaptive filtering resulted in a variety of algorithms, such as Frequency Bin Adaptive Filter (FBAF) [6, 7], correlation least square error (CLMS) algorithm [8], expanded CLMS [9, 10], double adaptation algorithm [11] with application in smart acoustic room (SAR) [12, 13], SAR system for car hands-free telephone [14], SAR for double-talk conditions [15], and others.

Applied Speech Processing
https://doi.org/10.1016/B978-0-12-823898-1.00008-4

© 2021 Elsevier Inc.
All rights reserved.

101

In addition to the proposed AHF, there are a variety of techniques to synthesize speech, for example, statistical parametric synthesis [16], hidden Markov models [17], deep neural networks [18], artificial neural networks [19], wavelet [20], wavenet [21], and others. Speech synthesis is used in different fields of intelligent speech processing [22] like speech recognition [23], speech tracking [24], augmentative interfaces [25], and others. The most popular application of speech synthesis is the text-to-speech system wherein an engine first transfers the text to linguistic symbols, then the speech synthesizer transfer the symbols to sound resembling speech [26]. The classic way to achieve this was by concatenating speech records from huge databases as needed. However, the prerecorded voice was subjected to various limitations in application, memory, and so on. Speech synthesizers produce speech without using any recordings.

A historical milestone in speech synthesis is Bell lab's vocoder, which extracted the fundamental tones from speech. It was later used by Dudley in 1938 to develop a voice demonstrator known as Voder [27]. Since then, advanced electronics and computers have helped rapidly progress speech synthesis, as shown in Table 5.1. Despite all these advances, synthetic speech is still clearly distinguishable from natural, real speech.

Table 5.1 Some historical works in speech synthesis.

Year	Pioneering works in speech synthesis
1779	A vocal tract sounding five vowels
1791	Acoustic mechanical speech machine
1837	Wheatstone speaking machine
1930	Bell lab's vocoder
1939	The Voder [27]
1940	Pattern playback [28]
1961	Kelly's voice (Bell lab) [29]
1968	Noriko Umeda's first general English text-to-speech system [30]
1966	Linear predictive coding introduced by Nippon Telegraph and Telephone (NTT)
1973	Texas Instruments LPC Speech Chips
1979	MUSA (MUltichannel Speaking Automaton) [31]
1995	DECtalk [32]
1997	Multilingual text-to-speech synthesis [33]
1980	The first video game to feature speech synthesis: "Shoot 'em up arcade game," Stratovox
1980	"Zero-cross" for synthesizing speech waveform [34]

5.2. Adaptive harmonic filtering approach to speech synthesis

Adaptive harmonic filtering (AHF) was initially proposed by Asharif et al. for estimation and cancellation of harmonic interference [1]. AHF adaptively generates the harmonic features of the signal by deploying adaptive learning; it obtains the sequence of harmonic components coefficients, which are the amplitudes of the signal tones. AHF is based on Fourier series and adaptive filter theory, as it deploys adaptive filtering techniques to follow the amplitudes of harmonic tones of the signal. As it known from Fourier series theorem, every harmonic signal is composed of sine and cosine waves weighted by different coefficients presenting as a linear combination known as Fourier series. Therefore, the signal under tracking is assumed as a harmonic signal, and AHF deploys an adaptive algorithm to adjust the coefficients of the equivalent Fourier series for the corresponding signal. The adaptive algorithm used by AHF does not use a weighted Tapped Delay Line, but instead generates the harmonic signal by internally weighting the sine and cosine components. Different adaptive learning algorithms can be used in AHF, like least square error (LMS), normalized least square error (NLMS), and proportional normalized least mean square (PNLMS).

Here, the theoretical frame of AHF is briefly explained. AHF internally uses an adaptation process in the Fourier series, which is adaptively made for the corresponding harmonic signal. As we know from Fourier series theory, any periodic signal can be expressed by superposition of a number of sine and cosine signals with different amplitudes, frequencies, and phases. In order to track the signal and synthesize the equivalent harmonic signal, the signal is assumed to be harmonic. In the case of speech this is not far from reality, since if we have a segment-wise view to the speech, each segment is a periodic signal, although it does not have a steady nature and changes dynamically. Considering $d(n)$ is the real speech signal for equivalent synthesis. $d(n)$ is periodic signal in a small, segment-wise view. Consider $y(n)$ as the synthetic speech that is going to be adaptively made. The sine and cosine components of $y(n)$ are considered as follows:

$$y_{\sin}(n) = \sin(i\omega_0 t)$$
$$= \sin\left(\frac{2i\pi f_0 n}{f_s}\right) \quad i = 1,2,...,N \tag{5.1}$$

$$y_{\cos}(n) = \cos(i\omega_0 t)$$
$$= \cos\left(\frac{2i\pi f_0 n}{f_s}\right) \quad i = 1,2,...,N \tag{5.2}$$

where $\omega_0 = 2\pi f_0$ is the considered fundamental frequency. $y(n)$ can be composed by superimposing the sine and cosine components:

$$y(n) = y_{\sin}(n) + y_{\cos}(n)$$
$$= \sum_{i=0}^{N} a_i(n) \sin\left(\frac{2i\pi f_0 n}{f_s}\right) + \sum_{i=0}^{N} b_i(n) \cos\left(\frac{2i\pi f_0 n}{f_s}\right) \tag{5.3}$$

The main effort of AHF is adaptively obtaining $a_i(n)$s and $b_i(n)$s by tracking $d(n)$, the targeted real speech, and minimizing the error between $y(n)$ and $d(n)$ using an adaptive algorithm like LMS. The LMS equations for the preceding AHF can be obtained as follows:

$$a_i(n+1) = a_i(n) + 2\mu e(n) \sin\left(\frac{2i\pi f_0 n}{f_s}\right)$$
$$b_i(n+1) = b_i(n) + 2\mu e(n) \cos\left(\frac{2i\pi f_0 n}{f_s}\right) \tag{5.4}$$

where $e(n)$ is the estimation error.

$$e(n) = d(n) - y(n) \tag{5.5}$$

Fig. 5.1 shows the structure of AHF, wherein updating coefficients are acquired separately for cosine and sine signals of different frequencies.

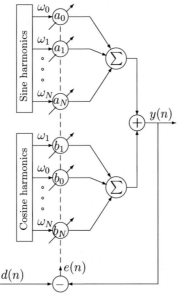

Fig. 5.1 Block diagram of an AHF.

To apply the AHF to generate a speech for the targeted real speech, it is enough to sequentially feed AHF by the real speech as $d(n)$, and initializing $a_i(n)$s and $b_i(n)$s as zero, as illustrated in Algorithm 5.1.

5.3. Experiments and results

In this section, we evaluate of efficiency the speech synthesis by AHF. The main quality of a synthetic speech is its level of similarity to the targeted human voice. This can be evaluated by objective numeral criteria like maximum cross correlation between the real speech segment uttering the same word and the synthetic one. Also, as a part of the Alan Turing test in evaluating artificial intelligence, the synthetic speech can be subjectively evaluated by human listeners as to the level of the ability to mimic a real human voice.

To perform the speech synthesis evaluation, we have applied AHF speech synthesis targeting nine different real speeches. These speeches are all from the same person saying in English "one," "two," "three," "four," "five," "six," "seven," "eight," and "nine." Fig. 5.2 shows the corresponding waves. For all of them, the AHF has been equivalently applied in 15,000 iterations, while the length of AHF for both cosine tabs and sine tabs is taken as 29 tabs. The adaptive learning step is acquired as $\mu = 0.003$. Fig. 5.3 shows

---

ALGORITHM 5.1 AHF in synthesizing $y(n)$ following the harmonic features of $d(n)$.

Input: $d(n)$ the real speech, μ step-size
Initialize: $[a_0, a_1, ..., a_N] = [0, 0, ...0]$;
$[b_0, b_1, ...b_N] = [0, 0, ...0]$;
while $n = n + 1$, *until error* $|e(n)|$ *reach a lower bound* **do**
  For $j = 0$ to N:

$$a_j(n + 1) = a_j(n) + 2\mu e(n)\sin\left(\frac{2j\pi f_0 n}{f_s}\right);$$
$$b_j(n + 1) = b_j(n) + 2\mu e(n)\cos\left(\frac{2j\pi f_0 n}{f_s}\right);$$
$$y(n) = \sum_{i=0}^{N} a_i(n)\sin\left(\frac{2i\pi f_0 n}{f_s}\right) + \sum_{i=0}^{N} b_i(n)\cos\left(\frac{2i\pi f_0 n}{f_s}\right);$$
$$e(n) = d(n) - y(n);$$

end
Output: $y(n)$;

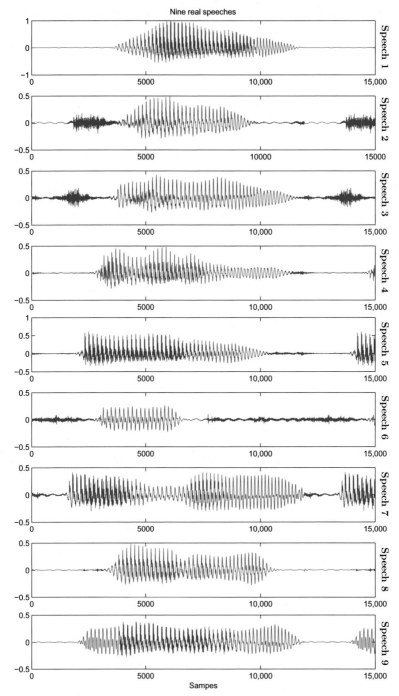

Fig. 5.2 Nine real speech segments (from *top* to *bottom*) saying "one," "two," "three," "four," "five," "six," "seven," "eight," and "nine."

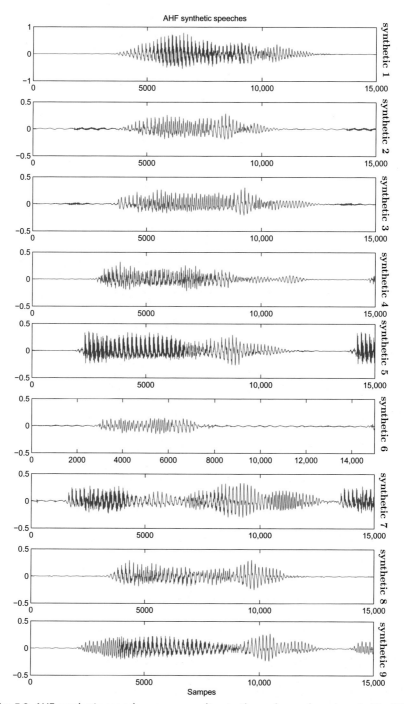

Fig. 5.3 AHF synthetic speeches corresponding to the real speeches given in Fig. 5.2.

the resultant synthetics speeches. The estimation error for each synthetic speech is calculated via Mean Squared Error in decibel (dB) as follows:

$$\text{Error (dB)} = 10\log_{10}\sum_{n=1}^{N}e^{2}(n) \qquad (5.6)$$

As the simulations are carried out via 10,000 Monte Carlo runs, in all cases, the AHF learning curve has a proper convergence, as shown in Fig. 5.4. The tuned sine and cosine tabs for each corresponding AHF are shown in Figs. 5.5 and 5.6, respectively.

As it is observable from the shape of the waves, there is a high level of similarity between morphology of each synthetic speech and its corresponding real one. As we listened to the adaptively harmonic synthesized speeches, not only did they sound very realistic and natural, but the voice was very clear and free of any noise. The noise-free melodic voice of the speech is the only aspect that makes synthetic speeches sound a bit different from real ones. This can be overcome by adding some natural noise to the speeches.

Apart from the personal subjective observation, we objectively evaluated the similarity of the synthetic and real speeches by normalized cross correlation (NCC) value among each couple of real and synthetic speeches. The formulation of NCC is as follows: NCC between two signals $x(n)$ and $\hat{x}(n)$, $n = 1, \dots, N$ is as follows:

$$\text{NCC} = \max_{n} \rho_{\hat{x}x}(n)$$

$$\rho_{\hat{x}x}(n) = \frac{r_{\hat{x}x}(n)}{\sqrt{r_{xx}(0)r_{\hat{x}\hat{x}}(0)}}$$

$$r_{\hat{x}x}(n) = \frac{1}{N}\sum_{k=0}^{N-n-1}x(k)\hat{x}(n+k) \qquad (5.7)$$

$$r_{xx}(0) = \frac{1}{N}\sum_{k=0}^{N-1}x(k)[x(k)]^{2}$$

$$r_{\hat{x}\hat{x}}(0) = \frac{1}{N}\sum_{k=0}^{N-1}x(k)\hat{a}(n+k)[\hat{x}(k)]^{2}$$

If we consider NCC as a criterion for identifying the corresponding real speech of each synthetic speech, the best way to judge is by making a confusion matrix according to nine available NCC values. Fig. 5.7 shows the confusion matrix. An observable from the diagonal pattern of the matrix,

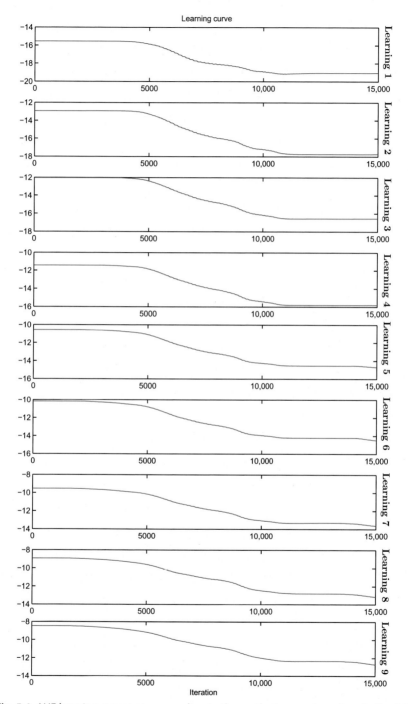

Fig. 5.4 AHF learning curves corresponding to the synthetic speeches given in Fig. 5.3.

Tuned sine tabs of AHFs

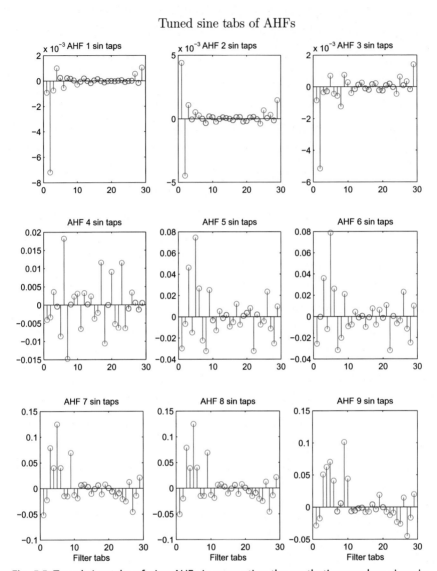

Fig. 5.5 Tuned sine tabs of nine AHFs in generating the synthetic speeches given in Fig. 5.3.

each synthetic signal shows maximum correlation with its corresponding real speech. Apart from that as listed in Table 5.2, all NCC values between the synthetic speeches and the corresponding real ones is more than 0.57, indicating a high level of similarity.

Tuned cosine tabs of AHFs

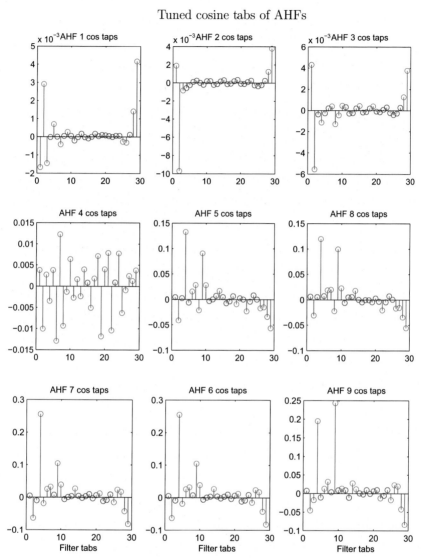

Fig. 5.6 Tuned cosine tabs of nine AHFs in generating the synthetic speeches given in Fig. 5.3.

5.4. Summary

This chapter presents a unique strategy for speech synthesis that adaptively forms harmonic components of speech and creates synthetic speech. This novel approach incorporates both adaptive filtering theorem and Fourier series. As speech can be segment-by-segment approximately presented by

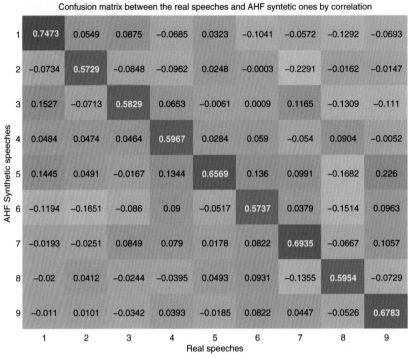

Fig. 5.7 Confusion matrix between real and synthetic speech using the normalized cross correlation.

Table 5.2 The NCC between each synthetic speech and its corresponding real speech.

Synthetic	One	Two	Three	Four	Five	Six	Seven	Eight	Nine
NCC	0.7473	0.5729	0.5829	0.5967	0.6569	0.5737	0.6935	0.5954	0.6783

Fourier series, a stationary speech segment is presentable by Fourier series. AHF tracks real speech and performs the online adjustment of the corresponding Fourier series. The objective and subjective evaluation of experiments over a number of real speeches proves the efficiency of AHF in synthesizing very natural-sounding synthetic speeches. The generated speeches show more than 57% correlation with their corresponding real speeches. In addition, AHF synthetic speeches sound very clear and natural. As such, the clear and melodic voice may not sound natural to the listener. In the future, we will add noise to synthetic speech to make it sound more natural.

References

[1] M.R. Asharif, N. Tomori, M. Khosravy, F. Asharif, K. Yamashita, T. Wada, Single sensor harmonic interference canceler (SSHIC), Proc. Signal Process. Symp. 27 (2012) 163–166.

[2] L. Zhang, M.R. Asharif, M. Khosravy, B. Senzio-Savino, F. Asharif, N. Itaru, Speech parameter generation algorithms for HMM-based speech synthesis, in: IEEJ 2017, Kyushu Fukumoto 2-1.

[3] M.R. Asharif, Asharif Digital Signal Processing Laboratory, University of the Ryukyus, Okinawa, Japan, 2020. https://ie.u-ryukyu.ac.jp/ asharif/Home/home.html (Last update March 2017).

[4] M.R. Asharif, K. Murano, M. Hatori, TV ghost cancelling by LMS-RAT digital filter, IEEE Trans. Consum. Electron. (4) (1981) 588–604.

[5] F. Amano, M.R. Asharif, S. Unagami, Y. Sakai, Echo canceller with short processing delay and decreased multiplication number, Google Patents, 1990. US Patent 4,951,269.

[6] M.R. Asharif, F. Amano, Acoustic echo-canceler using the FBAF algorithm, IEEE Trans. Commun. 42 (12) (1994) 3090–3094.

[7] M.R. Alsharif, Frequency domain noise canceller: frequency bin adaptive filtering (FBAF), in: Proc. IEEE, International Conference on Acoustic, Speech, and Signal Processing (ICASSP), 1986, pp. 2219–2222.

[8] M.R. Asharif, T. Hayashi, K. Yamashita, Correlation LMS algorithm and its application to double-talk echo cancelling, Electron. Lett. 35 (3) (1999) 194–195.

[9] M.R. Asharif, A. Shimabukuro, T. Hayashi, K. Yamashita, Expanded CLMS algorithm for double-talk echo cancelling, in: IEEE SMC'99 Conference Proceedings. 1999 IEEE International Conference on Systems, Man, and Cybernetics (Cat. No. 99CH37028), vol. 1, IEEE, 1999, pp. 998–1002.

[10] R. Chen, M.R. Alsharif, I.T. Ardekani, K. Yamashita, A new class of acoustic echo cancelling by using correlation LMS algorithm for double-talk condition, IEICE Trans. Fund. Spec. Section Issue Digital Signal Process. 87-A (8) (2004) 1933–1940.

[11] M.R. Asharif, L. Zhang, M. Khosravy, B. Senzio-Savino, F. Asharif, K. Yamashita, A new double adaptation algorithm for acoustic noise control, in: the 31st International Technical Conference on Circuits/Systems, Computers and Communications (ITCCSCC2016), July 10–13, IEICE Signal Processing Research Committee, 2016, pp. 701–704.

[12] R. Chen, F. Peng, Y. Luo, M.R. Alsharif, Smart acoustic room system and its application for car hand-free telephone, JSW 8 (2) (2013) 459–465.

[13] R. Chen, A.M. Reza, Smart acoustic room (SAR) system, in: the Proceedings of the Symposium on Information Theory and Its Applications, vol. 28, 2005, pp. 913–916.

[14] R. Chen, S. Kan, M.R. Alsharif, Echo and background music canceler by using smart acoustic room system for car hand-free telephone, in: 2009 First International Conference on Information Science and Engineering, IEEE, 2009, pp. 546–549.

[15] R. Chen, M.R. Alsharif, K. Yamashita, A new type echo cancelling by using the smart acoustic room (SAR) system & correlation function for the double-talk condition, in: IEEE, EURASIP, 9th International Workshop on Acoustic Echo and Noise Control (IWAENC 2005), 2005, pp. 29–32.

[16] H. Zen, K. Tokuda, A.W. Black, Statistical parametric speech synthesis, Speech Commun. 51 (11) (2009) 1039–1064.

[17] K. Tokuda, T. Yoshimura, T. Masuko, T. Kobayashi, T. Kitamura, Speech parameter generation algorithms for HMM-based speech synthesis, in: 2000 IEEE International Conference on Acoustics, Speech, and Signal Processing. Proceedings (Cat. No. 00CH37100), 3, IEEE, 2000, pp. 1315–1318. vol.

[18] H. Ze, A. Senior, M. Schuster, Statistical parametric speech synthesis using deep neural networks, in: 2013 IEEE International Conference on Acoustics, Speech and Signal Processing, IEEE, 2013, pp. 7962–7966.

[19] Z. Wu, O. Watts, S. King, Merlin: an open source neural network speech synthesis system, in: SSW, 2016, pp. 202–207.

[20] A. Suni, D. Aalto, T. Raitio, P. Alku, M. Vainio, Wavelets for intonation modeling in HMM speech synthesis, in: Eighth ISCA Workshop on Speech Synthesis, 2013.

[21] Oord Aaron, Yazhe Li, Igor Babuschkin, Karen Simonyan, Oriol Vinyals, Koray Kavukcuoglu, George Driessche, et al., Parallel wavenet: Fast high-fidelity speech synthesis, in: International conference on machine learning, PMLR, 2018, pp. 3918–3926.

[22] N. Dey, Intelligent Speech Signal Processing, Academic Press, 2019.

[23] S. Sen, A. Dutta, N. Dey, Audio Processing and Speech Recognition: Concepts, Techniques and Research Overviews, Springer, 2019.

[24] N. Dey, A.S. Ashour, Applied examples and applications of localization and tracking problem of multiple speech sources, in: Direction of Arrival Estimation and Localization of Multi-Speech Sources, Springer, 2018, pp. 35–48.

[25] B. Slotznick, Understanding phatic aspects of narrative when designing assistive and augmentative communication interfaces, Int. J. Ambient Comput. Intell. 6 (2) (2014) 75–94.

[26] J.P.H. Van Santen, R. Sproat, J. Olive, J. Hirschberg, Progress in Speech Synthesis, Springer Science & Business Media, 2013.

[27] H.W. Dudley, System for the artificial production of vocal or other sounds, 1938. US Patent 2,121,142.

[28] F.S. Cooper, A.M. Liberman, J.M. Borst, The interconversion of audible and visible patterns as a basis for research in the perception of speech, Proc. Natl. Acad. Sci. USA 37 (5) (1951) 318.

[29] B. Lambert, Louis Gerstman, 61, a specialist in speech disorders and processes, New York Times (1992). March 21.

[30] D.H. Klatt, Review of text-to-speech conversion for English, J. Acoust. Soc. Am. 82 (3) (1987) 737–793.

[31] L. Nebbia, P. Lucchini, Eight-channel digital speech synthesizer based on LPC techniques, in: ICASSP'79. IEEE International Conference on Acoustics, Speech, and Signal Processing, 4, IEEE, 1979, pp. 884–886. vol.

[32] W.I. Hallahan, DECtalk software: text-to-speech technology and implementation, Digit. Tech. J. 7 (4) (1995) 5–19.

[33] R.W. Sproat, Multilingual Text-to-Speech Synthesis, Kluwer Academic Publishers, 1997.

[34] J. Szczepaniak, The Untold History of Japanese Game Developers, vol. 3, SMG Szczepaniak, 2018.

PART 2

Speech identification, feature selection and classification

CHAPTER 6

Linguistically involved data-driven approach for Malayalam phoneme-to-viseme mapping

K.T. Bibish Kumar[a], Sunil John[a], K.M. Muraleedharan[a], and R.K. Sunil Kumar[b]
[a]Computer Speech & Intelligence Research Centre, Department of Physics, Government College, Madappally, Calicut, Kerala, India
[b]School of Information Science and Technology, Kannur University, Kannur, Kerala, India

6.1 Introduction

Speech, the natural mode of communication between human beings, is bimodal because it involves understanding both auditory and visual signals. Speech-based applications at public places like railway enquiry system, where clean speech is not available, is a pressing and current need. In such situations, visual signals are necessary for decision-making. For this, a bimodal speech processing technique is to be developed. However, processing the entire video signal is computationally expensive in any real-time speech processing application. Knowledge about visually separable atomic units in a language is a vital component in the development and design of any speech-based application.

A phoneme is the atomic sound unit necessary to symbolize all words in a particular speech. A viseme is the corresponding language unit of visual speech. For many years, the viseme has been studied in visual language and has undergone alterations to its definition. In the begining, a viseme is defined in terms of articulatory gestures such as mouth opening, teeth visibility, and tongue vulnerability that have to generate different phonemes, as in Ref. [1]. An equivalent definition that has been used extensively in the literature is a set of phonemes that has a similar visual look, as in Refs. [2, 3]. The static viseme does not account for the co-articulation effect of a visual address. The current description of a viseme is a lively visual language unit that describes distinct speech movements of the visual speech articulators, as in Ref. [4].

Applied Speech Processing
https://doi.org/10.1016/B978-0-12-823898-1.00003-5
© 2021 Elsevier Inc.
All rights reserved.
117

The contribution of the visual part in the judgment of speech, especially in a noisy environment, is vital. The visible organs of the articulatory system of human speech production consist of upper and lower lips, teeth, tongue, and lower jaw. The tongue, lips, and jaw are the most active visible articulators used in language production. Analysis of the lips, the most active visible articulator, is crucial in the visual speech analytics framework for speech recognition and synthesis.

Analysis of the visual speech signal and extracting the viseme set show satisfactory improvement in recognition of language components. A mapping between phoneme and viseme needs to be anticipated to synchronize the mouth form of distinct sounds. Before addressing the problems related to speaker variability [5], pose [6], choice of classifier technology [7–9], and recording device [10,11], the primary task in visual speech analysis is to decode the visual information from the lips. Since the extent of deformation of lips restricts as a result of facial muscles compared to the strain of the vocal organs, the viseme set in a speech is always smaller than the phoneme set. For developing the viseme set, the literature suggests two different approaches: linguistic and data-driven [12,13]. According to linguistic understanding, phonemes with the similar visual appearance of active articulators are treated as a viseme. In the data-driven strategy, visual speech is analyzed by extracting essential features from the lip area and grouping the features based on similarity measurement. The linguistic strategy is highly dependent on the perception ability of the linguistic or trained individual, which accurately represents the human lip-reading nature and is a time-consuming process. The data-driven approach provides a less time-consuming endeavor, but computational analysis of visual address profoundly depends on the choice of visual characteristics, which may be language-dependent. In human understanding, a holistic perspective is much more important than parts (Gestalt perception theory); however, in computer vision, parts (pixels) are much more significant than the whole (picture) [14]. As a result of this battle, the data-driven approach alone cannot mimic human perception accurately, and linguistic knowledge alone does not have the ability to examine substantial visual speech information. Therefore, a linguistically involved, data-driven approach can model individual perception from a linguistic approach with the computational ease of a data-driven approach.

The aim of this research work is to identify the viseme set in the Malayalam language using a linguistically involved, data-driven approach by consciously neglecting the issues in visual speech. One of the major problems in visual speech analysis is the unavailability of phonetically rich databases in the concerned language. Besides, there is no collective agreement in different aspects such as size of the database, the number of speakers, and the facial variability required to generalize the whole population for better improvement in the computational output. As an initial work in Malayalam visual speech analysis, we recorded an audio-visual speech database of 23 native speakers of Kerala by capturing the lip region of the speaker's face along with the audio. Malayalam is a syllable-based language written with a syllabic alphabet in which all consonants have an inherent vowel/a/. The database includes vowel and consonant-vowel syllables, which altogether comprises 50 phonemes.

The crucial step in visual speech analysis is identifying relevant static frames from the recorded video, which contains the visual appearance of the underlying phoneme. Based on the linguistic mapping, frames having a similar visual appearance were selected manually for each speaker, which minimized the error rate during data-driven analysis. The main problem to be addressed in the data-driven approach is the selection of relevant visual features to model human perception. This can be solved only by considering the visual speech properties of the underlined language. The tongue plays a vital role while uttering Malayalam speech, in terms of speed and flexibility, which makes it distinct from other languages. The degree of presence of teeth, oral cavity, and the shape of the lip using geometric features of lips and deformation in the appearance of lips and tongue can be modeled through the discrete cosine transform (DCT) feature. In this work, the mathematical analysis is initiated by extracting the geometric and DCT features from the selected frames.

The optimum number of static frames that represent the visual equivalent of a phoneme is an important factor to consider. Since the visual length and appearance of vowel phonemes and consonant phonemes are significantly different, static frames alone may end in sparse classification from the computational point of view. Besides, it is not advisable to consider every frame in a phoneme to represent the visual speech due to computational complexity and nonuniformity in the feature-length of vowel and consonant phonemes. To tackle this problem, time evolution of the static frame is implemented. In this work, a viseme is represented in three manners: static frame, static frame with preceding and following frames (three frames), and static frame with two preceding and following frames (five frames).

The next step is to identify the viseme set from visual features and the number of visemes. Researchers conducted a wide range of techniques to establish the mapping between phoneme and viseme, yet there is still no reliable and unambiguous method to confirm that one is better than the other. Within this work, we categorized the visual feature vectors to viseme groups by combining the K-means clustering method [15] and Gap statistic method [16]. Since K-means requires a predetermined cluster number, the Gap statistic method identifies the optimal cluster number from a range of clusters by exploring the language knowledge. The correlation between static visual speech unit and phoneme is carried out by developing a many-to-one phoneme-to-viseme mapping from isolated phonemes based on linguistic knowledge and by clustering in the parametric space using different visual features. Within this work, three viseme mappings (static frame alone, three frames, and five frames) are compared with the linguistic mapping and visual speech duration, thereby identifying the best representation of visual speech in terms of frames.

Through this work, a relevant methodology is developed for phoneme-to-viseme mapping. This methodology has an advantage that it considers the linguistic knowledge, which is an essential element in the creation of a speech-based application in the concerned language. A linguistically involved, data-driven approach can make an individual perception model from a linguistic approach with computational ease as it employs the data-driven approach as well. This is the first study that utilizes Gap statistics in the estimation of optimum cluster number from highly correlated visual speech data.

This chapter is organized as follows. Section 6.2 explains the concept of viseme mapping formation strategies. Section 6.3 describes the audio-visual Malayalam speech database developed for this work. Section 6.4 explains the different phoneme-to-viseme mappings derived using linguistic and parametric approaches. Section 6.5 discusses the durational analysis of visual speech and Section 6.6 presents the obtained results. Finally, Section 6.7 concludes with future directions.

6.2 Viseme set-formation approaches

Many researchers have analyzed the importance of the phoneme to viseme mapping. The phonemes that have almost the same mouth appearance are mapped to a single viseme class. In literature, many mappings are reported [17] where the number of visemes in a language varies between 10 and 20. The number and nature of visemes are language-dependent. Hence, a language-specific exploration is needed for establishing the viseme set for a particular language. Traditionally there are three approaches for obtaining visemes from a many-to-one mapping: linguistic knowledge-based [18–20], perception experiments with human subjects [1, 21, 22], and a data-driven approach [12,23–25]. Some authors blend the linguistic knowledge-based approach with perception experiments. This approach is termed as a subjective approach. In the subjective approach, viseme classes are defined through linguistic knowledge and prediction of phonemes having a similar visual appearance. A viseme class created by clustering of phonemes based on features extracted from the mouth region is the highlight of the data-driven approach. Most of the work in visual speech has been reported in European languages. There are few works reported in Indian languages such as Hindi [26–28] and Marathi [29]. Mishra et al. [26] developed a Hindi phoneme speech-recognition system using DCT as a visual feature and Mel frequency cepstral coefficient (MFCC) as an audio feature that reports better recognition in a noisy environment. Upadhyaya et al. [27] studied the performance of audio-visual speech-recognition systems under diverse noisy audio conditions using a different combination of image-based features. They studied the dependency of the recognition rate on different types of visual features and the nature of the acoustic noise used. Varshney et al. [28] clustered 23 phonemes into 5 viseme classes based on the data-driven approach using DCT as a visual feature. They also accomplished a viseme recognition task by integrating with audio features, which improved recognition rate. Brahme and Bhadade [29] presented phoneme viseme mapping for the Marathi language based on the linguistic approach alone. They derived 13 viseme classes, including silence from 44 phonemes. Attempts to identify viseme sets have not yet reached the realm of successful development of visual speech technology in Indian languages. This research work is the initial study in Malayalam phoneme-to-viseme mapping based on a linguistically involved, data-driven approach.

6.3 Malayalam audio-visual speech database

Malayalam, the official language of Kerala and the union territories of Lakshadweep and Puducherry, is an Indian language spoken by 40 million people. It is the youngest of the Dravidian language family and was designated as a classical language by the Government of India in 2013. Even though Malayalam developed double-rooted from Sanskrit and Tamil, it displays wide variation in another aspect, which makes Malayalam distinctive in Indian languages.

In this work, an audio-visual Malayalam speech database is created from 23 native speakers of Kerala (18 females and five males) by capturing the lip region of the speaker's face in normal lighting conditions along with the audio. The utterances are taken in the form of "silence-phoneme-silence" fashion. The language material of this database contains 10 vowel phonemes, two diphthongs, and 38 consonant-vowel syllables, which altogether comprise 50 phonemes, as in http://www.cmltemu.in/phonetic/#/.

6.3.1 Experimental setup

The recordings were made in a lab environment with natural daylight and ordinary illumination. Two handy cameras (high quality and low quality) were set to record the frontal view of the speaker. The cameras were mounted on a tripod stand about 100 cm away from the speaker's face. The low-quality camera was placed a little bit vertically above the high-quality camera. The low-quality camera was zoomed in to capture only the speaker's face, shoulder, and complex background. The high-quality camera was zoomed in to capture the mouth region, from chin to tip of the nose. Speakers were instructed to keep their heads in quasistatic condition and exhibit a neutral facial expression during recording. Speakers were allowed to practice elocution on the language material provided to them and modify their pronunciation. They were advised to widen their mouths so that the entire mouth region should be in the frame while uttering with such postures. Before recording, the camera screen was turned toward the speaker so that they could adjust their position for achieving the desired view in both cameras. Speakers were instructed to speak in their natural manner and were allowed out of the frame when they felt they made a mistake and let back in after rectifying it. The audio signals were captured using the inbuilt audio recorder in the two handy cameras in synchronization with the visual cues. The audio and video were both wrapped in MP4 files from the high-quality camera and MPG files from the low-quality camera. In this work, viseme mapping is carried out from the visual speech taken from high-quality cameras. A high-quality visual speech recorded from the speaker's mouth region had a resolution of 1280 × 720 and a frame rate of 25 fps in MP4 format. Two LED lamps with varying intensity ranging up to 400 W were used in the recording environment to capture the in-depth information from the speaker's mouth. The lamps were placed on both sides of the handy camera at equal distance from each other, thereby aligning all in a single line with the same vertical height. The light was slightly tilted toward the speaker's face, and the intensity adjusted to obtain uniform illumination on the speaker's face. This orientation was maintained throughout the recording process taken at different times. The approximate footage length is 5 min for isolated phonemes for each speaker. Fig. 6.1 shows the recording setup of the database.

6.3.2 Selection of relevant frames

After documenting the multimodal speech database, the visual speech mode alone was further examined for viseme mapping. The crucial step involved in the mapping procedure is the identification of relevant static frames from the recorded video, which contains the visual look of the underlying phoneme. This work is performed by rending the expertise from the linguistic peoples. In Malayalam, the language component in the audio speech (i.e., phoneme) is linguistically categorized according to articulation points and manners as in http://www.cmltemu.in/phonetic/#/. No such work is reported in Malayalam. Linguistic experts select the relevant frame based on the articulatory rules and use their knowledge to capture the dynamical variability of the speaker's mouth appearance. The first author of this work received training from linguistic experts for selecting the relevant frame. After choosing the frame for all phonemes of two speakers by the linguistic experts, this work was carried out to other speakers by the trained individual. The chosen frame also underwent a postselection endeavor by the coauthors of this work to minimize individual biasing and error rate during further analysis. To encode the time evolution of the visual speech of the underlying phoneme, two preceding and following frames from the chosen frame were selected. Fig. 6.2 shows the sequential arrangement of the frame, which

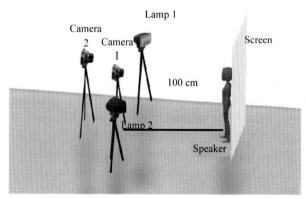

Fig. 6.1 Recording setup.

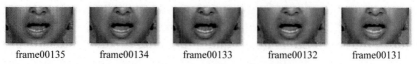

frame00135   frame00134   frame00133   frame00132   frame00131

Fig. 6.2 Sequential frames for phoneme -ଔ /a/.

captures the visual dynamics of the underlying phonemes. The linguistically chosen frame is the middle one, and two more frames selected from the preceding and following position of the selected one.

6.3.3 Lip contour extraction

The chosen frame may include areas/regions like background screen, hairs, ornaments, and so on that do not contribute anything to the visual speech and have the potential to induce serious influence in the analysis section. Before analyzing the linguistic and data-driven domain, undesirable areas are removed from the frames. Manual cropping cannot take care of the issues linked to the uniform framework area required to catch the dynamical variability in the appearance of underlying phoneme and speaker variability. In this work, we embraced a semiautomatic cropping method to deal with the issue and can further use it as a byproduct in the data-driven approach. The very first step in semiautomatic cropping is manual extraction of the lip shape information since the dynamicity related to the appearance of phoneme and speaker is embedded in the lip shape. To obtain the lip shape information, accurate lip contour is required. Bibish Kumar et al. [30] studied the lip region segmentation task using different color spaces in the database, which contains the entire complexion variability of Indian speakers. The insignificant variation in the color tone between the lip and the surrounding skin and the presence of facial hair creates a hindrance for capturing and analyzing lip dynamics automatically. Within this work, lip contour was obtained by fitting a straight line between the manually marked adjacent landmark points on each image. Lip contour was defined using 36 shape feature points on the lip. Twenty landmark points represent the outer lip contour, and 16 landmark points describe the inner lip contour. The number of landmark points was chosen by trial and error so that they best represent the lip contour. Since these points are equidistant, the contour is quasi smooth. Each landmark point contains an x coordinate and y coordinate, which is the pixel position in the image. Fig. 6.3 shows the images with lip contour marked manually using 36 landmark points. The second step in the semiautomatic cropping method is to automatically crop the frame using these landmark points. For this, the x

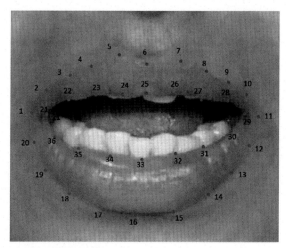

Fig. 6.3 Manual labeling landmark points in ROIs.

and y coordinates of the centroid of the lip contour ($X_{centoid}$ and $Y_{centoid}$) are estimated from the x and y coordinates of the landmark points (x_i and y_i) using (Eqs. 6.1, 6.2) for the centroid of a finite set of points. After this, a frame with dimension 600 × 500 whose mathematical center passes through the centroid of the lip shape of the original frame was cropped. The dimension of the frame was iteratively chosen to embed the lip region only.

$$X_{centoid} = \frac{1}{36} \sum_{i=1}^{36} x_i \tag{6.1}$$

$$Y_{centoid} = \frac{1}{36} \sum_{i=1}^{36} y_i \tag{6.2}$$

6.4 Malayalam phoneme-to-viseme/many-to-one mapping

In this work, linguistic and data-driven approaches were adopted for discovering the viseme set. The linguistic approach was carried out under the guidance of linguistic experts who analyzed the utterance style of the speakers. In the data-driven approach, the mathematical representation of visual speech was extracted and cluster based on the similarity measurement. Literature [31–33] shows a wide range of visual features based on the type of information embedded in them, including geometric-based, image-based, and model-based features. Geometric-based features explicitly simulate the measurement of mouth concerning height, weight, area, perimeter, and so on by analyzing the pixels in the lip boundary. Centroid distance and Fourier transcriptor [34] belong to this category. Image-based features consider all pixels in the region of interest (ROI) that are informative to represent the speech. In this method, the ROI is transformed into a different domain, thereby capturing the most informative components. Discrete cosine transform (DCT) [35,36], discrete wavelet transform (DWT) [37,38], principal component analysis (PCA) [39,40], linear discriminant analysis (LDA) [41], and combinations of these [42] belong to this category. Model-based features create a mathematical model to extract visual information with high computational complexity. Active shape model (ASM) and active appearance model (AAM) [43,44] belong to this class. Due to the diversity in lip movement of speakers, only language exploration can solve this issue. For uttering Malayalam sound, the tongue plays a vital role concerning flexibility and speed, which makes it distinct from other languages. The amount of teeth and oral cavity and the shape of the lip can be modeled with geometric features of lips, and deformation in the appearance of lips and tongue can be modeled through the DCT feature. The visual speech attributes are then clustered to identify the visual equivalent of the phoneme. Clustering is a vital step in data mining to discover the hidden pattern of an unlabeled dataset based on mathematical measurement. This method divides the dataset into smaller subclasses that have high intraclass similarity and low interclass similarity. The two widely used clustering algorithms are hierarchical (agglomerative & divisive) [45,46] and partitional (K-means) [15,47]. They explore the partition of data objects based on the number of clusters. However, the number of groups obtained from such approaches is highly sensitive to the nature of the dataset. Thus, identifying the optimal number of clusters is a significant endeavor, and can be carried out using the Gap statistic method [16,48]. To cluster a large and highly correlated dataset, K-means clustering, together with the Gap statistic method, is employed for optimum cluster selection in this work. The literature shows that this is the first work that employs the Gap statistic method in the development of viseme mapping.

6.4.1 Linguistic approach

In Malayalam, the language component in the audio speech (i.e., phoneme) is linguistically categorized according to articulation points and manners, as in http://www.cmltemu.in/phonetic/#/. The visual speech appearance depends primarily on the lip and lower jaw movements. Visibility of teeth and tongue is also a vital element. While uttering vowel phonemes, the lips are either wide open or projected outward. However, consonant phonemes are produced by touching the active articulator tongue at different places inside the mouth area whose dynamics is not visible. Thus the extent of appearance of active articulators is the discriminating factor to characterize the consonant phonemes. This section explores the possibilities of forming a viseme set from linguistic knowledge about the language and its phoneme set.

The vowel sound is generated by the flow of air from the larynx to the lips with no obstruction in the mouth region. Linguistically the five short vowel phonemes are distinguished by the position of the tongue as front-high, front-mid, central–low, back-high, and back-low. Since these tongue positions are partially visible, the shape information is utilized to categorize the visual equivalent of phonemes. Since the visual appearance of long vowels is the same as that of the corresponding short vowel, it is difficult to differentiate them visually. The front-high vowel ഇ /i/ exhibits a wide horizontal opening and the central-low vowel അ /a/ exhibits a vertical opening. Front-mid vowel എ /e/ is visually placed between front-high and central-low vowel. Both back-high ഉ /u/ and back-mid ഒ /o/ vowels show a rounded lip shape with a discriminating outward posture for the back-high vowel. The quick gliding of tongue from one vowel to another characterizes a diphthong. The diphthong ഐ /ai/ is made from the transition from അ /a/ to ഇ /i/ and ഔ /au/ is obtained from അ /a/ to ഉ /u/. The visual characterization of vowels is taken from the selected frame, whereas the visual signature of diphthongs is captured from the transition of the selected frames. Due to the diversity, two diphthongs are assigned to separate viseme classes. In short, each vowel is assigned to a separate viseme class, but monophthong short and long phonemes of the same vowel are placed in the same class. The viseme set for vowel and diphthong phonemes in Malayalam formed from linguistic understanding is given in Table 6.1.

In contrast to the vowel phoneme/sound, the consonant sound articulated with the complete or partial closure of the vocal tract is visually distinguishable only by considering the lip appearance. The most visually distinctive sound element in the consonant class is bilabial sound. While uttering this sound, the lips are kept closed with a slight strain in the facial muscles. The first consonant viseme classes formed from bilabial plosives expect ഫ /ph/. വ /va/, the only true labiodentals in Malayalam, and

Table 6.1 Linguistic classification of monophthong and diphthong vowel phonemes.

Viseme	Viseme class	Phoneme with IPA	Viseme in frame
1	Front, high—vowel	ഇ /i/, ഈ /iː/	
2	Front, mid—vowel	എ /e/, ഏ /eː/	
3	Central, low—vowel	അ /a/, ആ /aː/	
4	Back, high—vowel	ഉ /u/, ഊ /uː/	
5	Back, mid—vowel	ഒ /o/, ഓ /oː/	
6	Diphthong 1	ഐ /ai/	
7	Diphthong 2	ഔ /au/	

the bilabial plosive-voiceless aspirated ഫ /ph/, which is visually different from other Bilabial phonemes, are placed in the next viseme class due to the feeble presence of teeth. Viseme 10 consists of dental consonants, which have maximum teeth visibility and some traces of tongue tip. The velar consonants and the only glottal phoneme ഹ-ha is placed in the next viseme class since the place of articulation is back of the tongue and has the same visual appearance. The tongue is further backward in alveolar consonants, which are less visible and grouped in viseme class 12. Retroflex consonants are produced by curling the tongue backwardly and touching the front part of the hard palate, which produces the same visual appearance and thereby are assigned to a new viseme group. Viseme 14 is linguistically characterized as palatal consonants, which are produced by touching the tongue toward the hard palate. In brief, 50 isolated Malayalam phonemes were mapped into 14 viseme classes. The viseme set for the consonant phonemes in Malayalam formed from linguistic understanding is given in Table 6.2.

Table 6.2 Linguistic classification of consonant phonemes.

Viseme	Viseme class	Phoneme with IPA	Viseme in frame
8	Bilabial plosive-voiced and voiceless unaspirated, Nasal	പ്/p/, ബ് /b/, ഭ് /bh/, മ് /m/	
9	Bilabial plosive-voiceless aspirated and labiodental	ഫ് /ph/, വ് /v/	
10	Dental	ത് /t/, ഥ് /th/, ദ് /d/, ധ് /dh/, ന് /n̪/	
11	Velar Glottal	ക് /k/, ഖ് /kh/, ഗ് /g/, ഘ് /gh/, ങ് /ŋ/ ഹ് /h/	
12	Alveolar	റ്/r̠/, ന് /n/, സ് /s/, ര് /r/, റ് /r̠/, ല് /l/	
13	Retroflex	ട് /ʈ/, ഠ് /ʈh/, ഡ് /ɖ/, ഢ് /ɖh/, ണ് /ɳ/, ഷ് /ʂ/, ള് /ɭ/, ഴ് /ɻ/	
14	Palatal	ച് /c/, ഛ് /ch/, ജ് /ɟ/, ഝ് /ɟh/, ഞ് /ɲ/, ശ് /ʃ/, യ് /y/	

6.4.2 Data-driven approach

In data-driven approaches, visual features are extracted from the mouth region of talking faces and visemes are formed by clustering in the feature space. Both shape-based features and appearance-based features are used as visual cues in the Malayalam language. Shape-based features use information from the speaker's lip contour. The geometric features are the shape-based features used in this work. An appearance-based feature deals with pixel information in the ROI. As such it has high computational complexity and is weak in capturing geometric variations when compared to shape-based features. However, in a real-time application, appearance-based features show dominance over shape-based features, which have complexities related to accurate extraction of the lip contour. DCT is the appearance-based feature used in this work. Taking both methods together helps in judging their reliability in the problem under study. K-means clustering with the Gap statistic method is used to find the viseme set by clustering in the feature space.

6.4.2.1 Geometric visual features

Geometric features used in this study consist of outer lip width (w_{outer}), outer lip height (h_{outer}), inner lip width (w_{inner}), inner lip height (h_{inner}), outer lip area (a_{outer}), inner lip area (a_{inner}), and teeth area (t). These are extracted from the tracked lip contour as discussed in Section 6.2.

$$F_{geometric} = \{w_{outer}, h_{outer}, w_{inner}, h_{inner}, a_{outer}, a_{inner}, t_{area}\}.$$

The outer lip width and height are taken from the difference of x coordinate of landmark points 1 and 11 and y coordinate of landmark points 6 and 16, respectively, as in Fig. 6.4A. Similarly, the inner lip width and height are obtained from the difference of x coordinate of cardinal points 21 and 29 and y coordinate of cardinal points 25 and 33, respectively, as in Fig. 6.4B. The outer lip area is the total number of pixel points enclosed within the outer lip boundary as in Fig. 6.4C. The inner lip area is the oral cavity

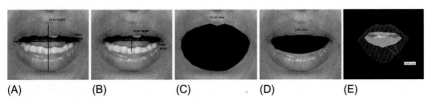

(A)   (B)   (C)   (D)   (E)

Fig. 6.4 Extraction of seven physical features from a frame.

region, which is measured by taking the total number of pixel points within the inner lip boundary as in Fig. 6.4D. The presence, absence, and area of teeth are direct indicators to distinguish many phonemes. The teeth area inside the convex hull of inner lip landmark points is computed after converting the pixels into the HSV color space [49]. Teeth pixels are segmented by using a thresholding process to the pixels inside the inner lip as in Fig. 6.4E.

6.4.2.2 Discrete cosine transform (DCT) visual features

DCT is one of the oldest and most popular appearance-based visual feature extraction techniques in the literature. A two-dimensional DCT of an M-by-N image is represented as:

$$D(i,j) = \sum_{i=1}^{M} \sum_{j=1}^{N} I(i,j) \cos\left(\frac{(2i+1)\pi i}{2M}\right) \cos\left(\frac{(2j+1)\pi J}{2N}\right) \quad (6.3)$$

where $I(i, j)$ is the gray-scale image of the ROI. The DCT return a two-dimensional matrix having $M * N$ coefficients. Most of the visually significant information and energy is concentrated in a few coefficients of DCT, which represent the low-frequency aspect of an image. Initially, the cropped speech frame of size 600×500 is further reduced to 64×64 for better implementation of the DCT algorithm. To avoid the curse of dimensionality, first, 20 coefficients per frame are selected from a 64×64 DCT coefficients in a zig-zag manner [50] starting from DC component ($D(1,1)$).

6.4.2.3 Viseme set formation by clustering in the parametric space

The viseme set is formed by clustering in the feature vector space. Shape-based and appearance-based visual feature vectors are analyzed separately for 50 whole phonemes and 38 consonant phonemes. The geometric feature comprises 7 numerical values per frame, and DCT features comprise 20 numerical values per frame. Thus each phoneme is represented by a 35-dimensional geometric feature vector and 100-dimensional DCT feature vector, respectively. The numerical representation of each phoneme is arranged horizontally in the feature vector. An aggregate feature vector is created by horizontally concatenating the feature vector of 23 speakers. This feature vector was standardized for further analysis.

The final feature vector is fed into the Gap statistic method for determining optimum cluster number by making use of the K-means algorithm for clustering purposes. The K-means algorithm is one of the simplest unsupervised learning algorithms that classifies the dataset into a predefined number of clusters based on the centroid, as in Ref. [47]. The algorithm inputs are the dataset containing "n" objects and predefined cluster number "k." The algorithm of K-means clustering is given below.

1. The algorithms start with initial estimates for the K centroids, which can either be randomly generated or randomly selected from the data set.
2. Each centroid defines one of the clusters. In this step, each data point is assigned to its nearest centroid, based on the squared Euclidean distance.
3. In this step, the centroids are recomputed by taking the mean of all data points assigned to that centroid's cluster.
4. The algorithm iterates between steps two and three until a stopping criterion is met (i.e., no data points change clusters, the sum of the distances minimized, or some maximum number of iterations reached).

Due to the high correlation of mouth parameters for acoustically different phonemes, the clustering algorithm alone fails to estimate an optimum cluster value. The Gap statistic method has proven its strength in identifying optimal cluster numbers in an extremely correlated dataset. The Gap statistic method compares the intraclass dispersion obtained from the given data with that of an appropriate reference distribution (Tibshirani, 2001). The methodology of Gap statistic is explained in the paragraph that follows (using the notation from Tibshirani, 2001).

Consider a dataset $\{x_{ij}\}$ with $i = 1, 2, \ldots, n$ and $j = 1, 2, \ldots, p$, consists of p features measured on n independent observations, clustered into k clusters $C_1, C_2, \ldots, C_k$, where C_r denotes the indexes of samples in cluster r, and $n_r = |C_r|$. Let $d_{ii'}$ denotes the squared Euclidean distance between the observation i and i' $(d_{ii'}, \sum_j (x_{ij} - x_{i'j})^2)$. The sum of the pairwise distance D_r for all points in cluster r is:

$$D_r = \sum_{i, i' \in Cr} d_{ii'} \tag{6.4}$$

Let W_k be the within–cluster sum of squared distances from the cluster means as:

$$W_k = \sum_{r=1}^{k} \frac{1}{2n_r} D_r \tag{6.5}$$

W_k decreases monotonically as the number of clusters k increases. For calculation, the Gap function, Tibshirani et al. [48] proposed to use the difference of the expected value of $\log(W^*_k)$ of an appropriate null reference and the $\log(W_k)$ of the dataset,

$$\text{Gap}_n(k) = E^*_n \log(W^*_k) - \log(W_k) \qquad (6.6)$$

where E^*_n denotes the expectation of under a sample of size n from the reference distribution. Then the proper number of clusters for the given data is the smallest k such that.

$$\text{Gap}_n(k) \geq \text{Gap}_n(k+1) - s_{k+1} \qquad (6.7)$$

Let s_k be the simulation error calculated from the standard deviation $sd(k)$ of B Monte Carlo replicates $\log(W^*_k)$ according to the equation $s_k = \sqrt{1 + 1/B}\, sd(k)$, which is represented by a vertical bar in Gap curve.

In short, a range of cluster groups is estimated using the K-means algorithm (or any other clustering algorithm), and the logarithm of within-cluster variance is compared with the same measurement of an appropriate reference distribution of the data. The difference between these quantities (Gap curve or Gap function) provides the corresponding Gap value for each cluster group, and these clusters show a fall at the point where the Gap is maximum. Fig. 6.4 reveals different Gap statistic steps of data using K-means clustering. For a well-separated dataset, the Gap function exhibits a monotone behavior as in Fig. 6.5. However, for a highly correlated dataset, the Gap function exhibits a nonmonotone behavior, which directly educates us to inspect the whole Gap curve instead of simply finding the cluster number with the maximum Gap. The performance of the Gap statistic profoundly depends on the nature of the database and feature vector used. By considering the problem under study, features should select in such a way that it can display a high discriminating power between high correlated observables. In this work, the redundancy of selected features has been removed by considering a few feature coefficients that may deal with issues under study. The optimum number of features required to deal with the issue under study is still an open research area.

Depending upon the property of the dataset and the underlying problem, it is better to study a reasonable range in the Gap curve. Since analyzing the whole gap curve is tiresome and time-consuming in estimating the optimum cluster number, especially for highly correlated data. For a highly correlated dataset and phoneme-to-viseme conversion problem, it is better to examine the Gap curve between the clusters of 10–20. Though the amount of viseme is language-dependent, the majority of the published works have underlined this range in various languages. In this work as well, 50 Malayalam phonemes are linguistically mapped to 14 viseme classes. For a straightforward interpretation, a minimum of two phonemes can occupy a single viseme class due to

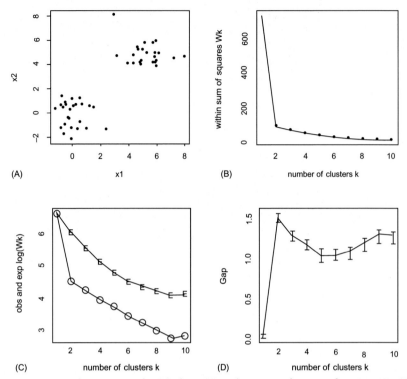

Fig. 6.5 A two-cluster example: (A) data; (B) within sum of square function W_k; (C) functions $\log(W_k)$ (O) and $E *_n\{\log(W_k)\}$ (E); (D) gap curve. *(Courtesy: Tibshirani (2001).)*

high correlation in the visual appearance of phonemes, thereby creating a set of 25 visemes. The same methodology is carried out in the rest of this work. For clustering the feature vectors of 50 phonemes, Gap curve analyzed, and the optimum cluster number was identified with maximum Gap value in the cluster range 10–20, as shown in Fig. 6.6. For 50 Malayalam phonemes, based on geometric feature vector and DCT feature vector using the static frame alone, the estimated viseme set is 16 and shown in Table 6.3.

One of the most important inferences from Table 6.3 is the visual appearance of the vowel phoneme is almost embedded in a single frame. In addition, the velar consonant phonemes (ക്-/k/, ഖ്-/kh/, ഗ്-/g/, ഘ്-/gh/, ങ്-/ŋ/) were distinguished from other consonant groups, just like in the linguistic mapping. All other consonant groups were randomly distributed while analyzing the viseme using a single frame. The viseme map using three frames is shown in Table 6.4.

While taking the preceding and following frames, the majority of the consonant phonemes were almost matched with Tables 6.1 and 6.2. Along

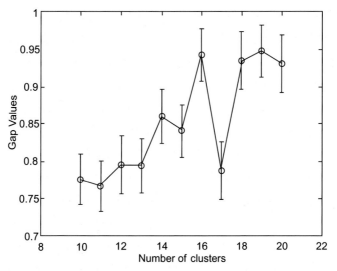

Fig. 6.6 Gap curve.

Table. 6.3 Phoneme-to-viseme mapping using one frame based on the geometric and DCT feature vectors.

Viseme	Phonemes with IPA—geometric feature vector	Phonemes with IPA—DCT feature vector
1	ച്-/c/, ഛ്-/cʰ/, ജ്-/ɟ/, ഡ്-/ ɟʰ/, ഞ്-/ɲ/, ത്-/t/, ഥ്-/tʰ/, ദ്-/d/, ധ്-/dʰ/, ന്-/n̪/, ശ്-/ʃ/, ഷ്-/ʂ/, സ്–/s/, റ്റ്–/ṟ/	ഉ-/u/, ഊ-/uː/
2	ബ്-/b/, മ്-/m/	ല്-/l/
3	ഹ്-/h/, ള്-/ɭ/	ഓ-/o/, ഓാ-/oː/
4	യ്-/y/, ര്-/r/, ല്-/l/, ഴ്-/ɹ/, റ്-/ṟ/, ന്-/n/	ച് -/c/, ഛ്-/cʰ/, ജ്-/ɟ/, ഡ്-/ɟʰ/, ഞ്-/ɲ/
5	അ-/a/, ആ-/aː/	ഓാ-/au/
6	ഉ-/u/, ഊ-/uː/, ഓ-/o/, ഓാ-/ o:/	ക്-/k/, ഖ്-/kʰ/, ഗ്-/g/, ഘ്-/gʰ/, ങ്-/ŋ/, ണ്-/ɳ/
7	ഭ്-/bʰ/	ള്-/ɭ/, ര്-/ṟ/
8	പ്-/p/	എ)-/e/, ഏ)-/eː/, ഐ)-/ai/
9	എ)-/e/, ഏ)-/e:/	ര്-/ṟ/

Continued

Table. 6.3 Phoneme-to-viseme mapping using one frame based on the geometric and DCT feature vectors—cont'd

Viseme	Phonemes with IPA—geometric feature vector	Phonemes with IPA—DCT feature vector
10	ഫ്-/pʰ/, വ്-/v/	ത്-/t/, ഥ്-/tʰ/, ദ്-/d/, ധ്-/dʰ/, ന്-/n̪/, ഠ്-/ʈʰ/, ഡ്-/ɖ/, ഢ്-/ɖʰ/, ഫ്-/pʰ/, വ്-/v/
11	ഇ-/i/, ഈ-/i:/	ട്-/ʈ/, യ്-/y/, ര്-/r/, ഴ്-/ɻ/, ന്-/n/
12	ഠ്-/ʈʰ/, ഢ്-/ɖʰ/	പ്-/p/, ബ്-/b/, ഭ്-/bʰ/, മ്-/m/
13	ഔ-/au/,	സ്-/s/
14	ഐ-/ai/	അ-/a/, ആ-/a:/, ഹ്-/h/
15	ട്-/ʈ/, ഡ്-/ɖ/, ണ്-/ɳ/	ശ്-/ʃ/, ഷ്-/ʂ/
16	ക്-/k/, ഖ്-/kh/, ഗ്-/g/, ഘ്-/gh/, ങ്-/ŋ/	ഇ-/i/, ഈ-/i:/

Table 6.4 Phoneme-to-viseme mapping using three frames based on the geometric and DCT feature vectors.

Viseme	Phonemes with IPA—geometric feature vector	Phonemes with IPA—DCT feature vector
1	ക്-/k/, ഖ്-/kʰ/, ഗ്-/g/, ഘ്-/gʰ/, ങ്-/ŋ/	പ്-/p/, ബ്-/b/, ഭ്-/bʰ/, മ്-/m/
2	ഓ-/o:/	എ)-/e/, ഏ-/e:/
3	പ്-/p/, ബ്-/b/, ഭ്-/bʰ/, മ്-/m/	ഐ-/ai/
4	ട്-/ʈ/, ഠ്-/ʈʰ/, ഡ്-/ɖ/, ഢ്-/ɖʰ/, ണ്-/ɳ/	ഇ-/i/, ഈ-/i:/
5	ഉ-/u/, ഊ-/u:/	ശ്-/ʃ/, ഷ്-/ʂ/, സ്-/s/, റ്-/r̠/
6	ഫ്-/pʰ/	ഛ്-/cʰ/, ജ്-/ɟ/, ഝ്-/ɟʰ/
7	യ്-/y/, ര്-/r/, ല്-/l/, ഹ്-/h/, ള്-/ɭ/, ഴ്-/ɻ/, റ്-/ɾ/, ന്-/n/	ഒ-/o/, ഉ-/u/, ഊ-/u:/
8	ച്-/c/, ഞ്-/ɲ/	അ-/a/, ആ-/a:/, ഹ്-/h/
9	എ)-/e/, ഏ-/e:/, ഐ-/ai/	ഫ്-/pʰ/, വ്-/v/

Table 6.4 Phoneme-to-viseme mapping using three frames based on the geometric and DCT feature vectors—cont'd

Viseme	Phonemes with IPA—geometric feature vector	Phonemes with IPA—DCT feature vector
10	ഔ-/au/	ച്-/c/, ഞ്-/ɲ/
11	ഇ-/i/, ഈ-/iː/, ത്-/t/, ഥ്-/tʰ/, ദ്-/d/, ധ്-/dʰ/, ന്-/n̪/	ര്-/r/, ല്-/l/, ള്-/ɭ/, റ്-/ɾ/, ഴ്-/ʐ/, ന്-/n/
12	ശ്-/ʃ/, ഷ്-/ʂ/, സ്-/s/, റ്-/ɾ/	ട്-/ʈ/, /, ഠ്-/ʈʰ/, ഡ്-/ɖ/, ഢ്-/ɖʰ/, ണ്-/ɳ/ ത്-/t/, ഥ്-/tʰ/, ദ്-/d/, ധ്-/dʰ/, ന്-/n̪/
13	വ്-/v/	ഓ-/oː/, ഔ-/au/
14	ഛ്-/cʰ/, ജ്-/ɟ/, ഝ്-/ɟʰ/	യ്-/y/
15	അ-/a/, ആ-/aː/	ക്-/k/, ഖ്-/kʰ/, ഗ്-/g/, ഘ്-/gʰ/, ങ്-/ŋ/
16	ഒ-/o/	

with the velar consonant phonemes, bilabial consonant phonemes (പ്-/p/, ബ്-/b/, ഭ്-/bʰ/, മ്-/m/) and labiodental consonant phonemes (ഫ്-/pʰ/, വ്-/v/) were separated into distinct viseme classes in both features. While comparing Tables 6.4 and 6.5, the DCT feature vector has shown closer resemblance to linguistic mapping than the geometric features have. Viseme representation using five frames is shown in Table 6.5.

Based on geometric features, the diphthong vowel phonemes (ഔ-/au/), central vowel phonemes (അ-/a/, ആ-/aː/), bilabial consonant phonemes (പ്-/P/, ബ്-/b/, ഭ്-/bh/, മ്-/m/), labiodental consonant phonemes (ഫ്-/pʰ/, വ്-/v/), dental consonant phonemes (ത്-/t/, ഥ്-/tʰ/, ദ്-/d/, ധ്-/dʰ/, ന്-/n̪/) velar consonant phonemes (ക്-/k/, ഖ്-/kʰ/, ഗ്-/g/, ഘ്-/gʰ/, ങ്-/ŋ/), and most of the palatal consonant phonemes (ച്-/c/, ഛ്-/cʰ/, ജ്-/ɟ/, ഝ്-/ɟʰ/, ഞ്-/ɲ/)are grouped exactly in the same manner as in the linguistic approach (Tables 6.1 and 6.2). Due to lip contour similarity among front-high vowels (ഇ-/i/,ഈ-/iː/) and front-mid vowels (എ-/e/,ഏ-/eː/) and back-high vowels (ഉ-/u/,ഊ-/uː/) and back-mid vowel (ഒ-/o/, ഓ-/oː/), they are selected to the same viseme class rather than different classes, as in the linguistic approach. The remaining consonant phonemes are distributed in such a way that it follows some traces of the linguistic point of the consonant phoneme cluster.

Table 6.5 Phoneme-to-viseme mapping using five frames based on the geometric and DCT feature vectors.

Viseme	Phonemes with IPA—geometric feature vector	Phonemes with IPA—DCT feature vector
1	റ്-/r̠/	ഒ-/o/, ഓ-/o:/
2	ഉ-/u/,ഊ-/u:/,ഒ-/o/, ഓ-/o:/	ക്-/k/, ഖ്-/k^h/, ഗ്-/g/ , ഘ്-/g^h/, ങ്-/ŋ/
3	അ-/a/, ആ-/a:/, ഹ്-/h/	ഉ-/u/, ഊ-/u:/
4	പ്-/p/,ബ്-/b/,ഭ്-/b^h/,മ്-/m/	പ്-/p/, ബ്-/b/, ഭ്-/b^h/, മ്-/m/
5	ല്-/l/, ള്-/ɭ/	അ-/a/, ആ-/a:/, ഹ്-/h/
6	ഫ്-/p^h/	ച് -/c/, ഛ്-/c^h/, ജ്-/ɟ/, ഝ്-/ɟ^h/, ഞ്-/ɲ/
7	ത്-/t/, ഥ്-/t^h/, ദ്-/d/, ധ്-/d^h/, ന്-/n̪/	എ-/e/, ഏ-/e:/, ഐ-/ai/
8	യ്-/y/,ര്-/r/,ന്-/n/	ഇ-/i/, ഈ-/i:/
9	ച്-/c/, ഛ്-/c^h/, ജ്-/ɟ/, ഝ്-/ɟ^h/, ഞ്-/ɲ/	ട്-/ʈ/, ഠ്-/ʈ^h/, ഡ്-/ɖ/, ഢ്-/ɖ^h/, ണ്-/ɳ/,
10	റ്-/ṛ/	റ്-/r̠/, ന്-/n/
11	ക്-/k/,ഖ്-/k^h/, ഗ്-/g/, ഘ്-/g^h/, ങ്-/ŋ/	ഔ-/au/
12	വ്-/v/	ശ്-/ʃ/, ഷ്-/ʂ/, സ്-/s/,
13	ട്-/ʈ/, ഠ്-/ʈ^h/, ഡ്-/ɖ/, ഢ്-/ɖ^h/, ണ്-/ɳ/, ശ്-/ʃ /, ഷ്-/ʂ/, സ്-/s/, ഴ്-/ʐ/	യ്-/y/, ര്-/r/,ല്-/l/,
14	ഔ-/au/	ത്-/t/, ഥ്-/t^h/, ദ്-/d/, ധ്-/d^h/, ന്-/n̪/
15	ഇ-/i/, ഈ-/i:/, എ-/e/, ഏ-/e:/, ഐ-/ai/	ള്-/ɭ/, ഴ്-/ʐ/, റ്-/ṛ/
16		ഫ്-/ph/, വ്-/v/

Phoneme-to-viseme mapping was also studied for DCT feature vectors with an estimated optimal cluster number equal to 16. The vowel phonemes are distributed precisely in the same fashion as in the linguistic approach, but with overlapping of a diphthong phoneme (ഐ-/ai/) into the front-mid vowel phoneme class (എ-/e/, ഏ-/e:/). Most of the consonant phonemes distribute precisely as in the linguistic approach. The remaining consonant phonemes are randomly distributed, just like in geometric feature maps.

In addition, the authors studied the viseme map using seven frames; however, no noticeable variation was observed from the viseme map using five frames. All vowel phonemes can be modeled with a single frame, but for consonant phonemes, nearly five frames are needed to represent their visual appearance. This may be because the visual appearance of the articulators involved in pronunciation of consonant phonemes are feebler than vowel phonemes, which mainly need the lip shape information. The next section deals with the durational analysis of visual speech, which justifies the requirement of five frames to visually represent the consonant phonemes.

6.5 Durational analysis of visual speech

Time annotation of the speech database is a time-consuming and error-prone task. In this work, the durational analysis is carried out on the speech database of five speakers uttering all phonemes. Manual segmentation was guided by auditory perception. Each phoneme was presented in isolation, and the phoneme boundaries were adjusted so that the audible traces of preceding or following phonemes were minimized. To minimize individual biasing and error rate, the segmented phonemes were rechecked with another listener. In the speech, there is an inherent asynchrony between the visual and audio cues, but in this work, only the visual speech signal is studied. The detailed description of the segmentation process of the same database is explained in Bibish Kumar et al. [51].

Fig. 6.7 shows the visual duration of all Malayalam phonemes with duration in seconds along the y-axis and phonemes along the x-axis. The triangle in the stock plot represents the mean duration of the corresponding phoneme and the length of the line from the top and bottom of the triangle indicates "mean + standard deviation" and "mean − standard deviation," respectively. The duration of vowel phonemes ranges between 0.50 and 0.70 s, which corresponds to 12 and 17 frames, respectively (frame rate of the video is 25 Hz). The duration of consonant phonemes varies between 0.10 and 0.30 s, which corresponds to two and seven frames, respectively. The average duration of Malayalam vowel and consonant phonemes from the visual speech is 0.606 and 0.200 s, which corresponds to 15 and 5 frames, respectively. Hence it is better to represent the visual equivalent of a phoneme with the time evolution of a linguistically identified frame of duration 0.20 s, more precisely, preceding and proceeding two frames from the selected frame.

As shown in Table 6.5, the consonant phonemes were distributed in the same manner of the linguistic picture when the visual speech is represented

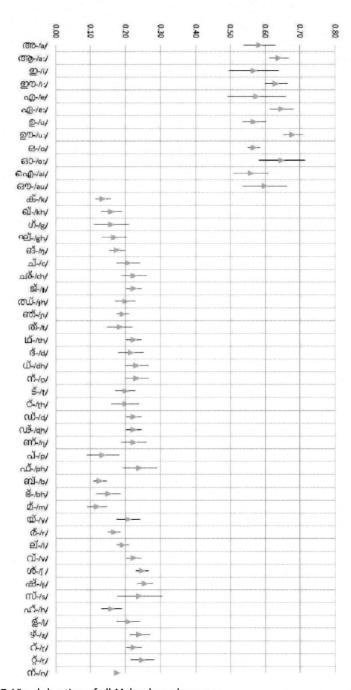

Fig. 6.7 Visual duration of all Malayalam phonemes.

in five frames, which are exactly shown in the durational analysis as well. Thus for the Malayalam language, a relevant frame with preceding two and following two frames (total five frames) is necessary to model the visual speech element, especially the visual equivalent of the consonant phoneme.

6.6 Discussion

In this work, we introduced a linguistically involved, data-driven approach for the viseme set identification problem. Automatic identification of visemes (frames) for corresponding underlying phonemes from visual speech is still an open research area. In a computational point of view, a linguistic approach alone does not have the ability to process massive visual speech information. In this work, we defined the viseme as the time evolution of a linguistically selected frame for the underlying phoneme. For this, three phoneme-to-viseme mappings were developed using a linguistically selected frame, the selected frame along with the previous and following frames, and the selected frame with two previous and two following frames. The strength of visual speech features (geometric-based and image-based) was also evaluated by comparing the corresponding viseme map with the linguistic map.

The vowel phonemes alone displayed almost similar grouping, even when the viseme is represented by a single frame, as in Table 6.3. However, the classification becomes worse for viseme mapping based on geometric features as in Table 6.5. However, a viseme map based on DCT features shows strength in classifying the vowel phonemes in all three formats of the viseme. Hence the visual appearance of the vowel phonemes can be modeled even with a single frame based on both features. However, geometric features reduce strength when the viseme is represented by the time evolution of the selected frame. For consonant phoneme grouping, viseme maps using five frames (Table 6.5) showed close resemblance with the linguistic map as in Table 6.2. Hence the visual appearance of the consonant phoneme can be modeled only by considering the time evolution of the linguistically identified frame. In terms of features, DCT features are always dominant over geometric features in each viseme mapping. Besides, the complexity of extraction of exact lip contour in geometric features both automatically and manually (in this work) makes the image-based features (here, DCT) more prior in the visual speech extraction process.

6.7 Conclusion

In this chapter, we presented a novel approach to phoneme-to-viseme mapping in the Malayalam language and identifying the number of frames needed to model the visual speech element using viseme mapping and visual durational analysis. We used a linguistically involved data-driven approach for the viseme mapping. An audio-visual Malayalam speech database consisting of 23 trained speakers uttering 50 isolated Malayalam was used for this work. In the first phase of the study, phoneme-to-viseme many-to-one mapping was done according to the linguistic and data-driven approaches. In the linguistic approach, 50 phonemes were grouped into 14 viseme classes based on linguistic knowledge. In the data-driven approach, three viseme sets were created by clustering the numerical representation of phonemes using the K-means algorithm and Gap statistic. Geometric and DCT visual features were used in the data-driven approach. The DCT features-based viseme map showed close resemblance with the linguistic map by classifying 50 phonemes into 16 viseme classes. The visual speech element equivalent to vowel phonemes can be modeled using a relevant frame that represents the corresponding phoneme. For consonant phoneme, five consecutive frames are needed to model the equivalent visual speech.

In the second phase, the requirement of five frames to visually represent the consonant phonemes was verified by analyzing the visual duration of all Malayalam phonemes. The average durational range of vowel phonemes is 0.5–0.7 s, which corresponds to 12 frames to 17 frames. Since the vowel phonemes can be visually modeled using a relevant single frame, additional frames may increase the computational cost. The average durational range of consonant phonemes is 0.1 and 0.3 s, which corresponds to two frames to seven frames. Thus for Malayalam language, five consecutive frames are required to represent the visual speech element, especially the visual equivalent of the consonant phonemes.

This present work is the result of efforts taken in the past 2 years: for database creation, manual lip contour extraction, and time alignment of our database. The performance of our methodology heavily depends on the selection of relevant frames of the corresponding phoneme using linguistic knowledge, which is a time-consuming process. Only a stochastic model is able to capture the facial movement from a talking face. Developing these maps based on supervised learning algorithms has potential to improve the present viseme set.

Acknowledgements

This work is a revision of our previous work and contains a significant portion thereof. Journal article: Bibish Kumar, K.T., Sunil Kumar, R.K., Sandesh, E.P.A., et al., 2019. Viseme set identification from Malayalam phonemes and allophones. Int. J. Speech Technol. 22, 1149–1166. https://doi.org/10.1007/s10772-019-09655-0.

References

[1] C.G. Fisher, Confusions among visually perceived consonants, J. Speech Hear. Res. 11 (4) (1968) 796–804.

[2] H.L. Bear, R. Harvey, Decoding visemes: improving machine lip-reading, in: H.L. Bear, R. Harvey (Eds.), IEEE International Conference on Acoustics, Speech and Signal Processing (ICASSP), 2016, 2016, pp. 2009–2013.

[3] E. Bozkurt, Ç.E. Erdem, E. Erzin, T. Erdem, M. Özkan, Comparison of phoneme and viseme based acoustic units for speech driven realistic lip animation, in: Proceedings of 3DTV-CON, 2007, https://doi.org/10.1109/3DTV.2007.4379417.

[4] S.L. Taylor, M. Mahler, B.J. Theobald, I. Matthews, Dynamic units of visual speech, in: Computer Animation 2012—ACM SIGGRAPH/Eurographics Symposium Proceedings, SCA 2012, 2012, pp. 275–284.

[5] H.L. Bear, R. Harvey, Comparing heterogeneous visual gestures for measuring the diversity of visual speech signals, Comput. Speech Lang. 52 (2018) 165–190, https://doi.org/10.1016/j.csl.2018.05.001.

[6] P. Lucey, G. Potamianos, Lipreading using profile versus frontal views, in: 2006 IEEE 8th Workshop on Multimedia Signal Processing, MMSP 2006, 2007, pp. 24–28, https://doi.org/10.1109/MMSP.2006.285261.

[7] K. Noda, Y. Yamaguchi, K. Nakadai, H.G. Okuno, T. Ogata, Audio-visual speech recognition using deep learning, Appl. Intell. 42 (4) (2015) 722–737, https://doi.org/10.1007/s10489-014-0629-7.

[8] M. Sarma, K.K. Sarma, Recent Trends in Intelligent and Emerging Systems, May 2015, 2015, pp. 173–187.

[9] S. Sen, A. Dutta, N. Dey, Audio Processing and Speech Recognition: Concepts, Techniques and Research Overviews, 2019, https://doi.org/10.1007/978-981-13-6098-5.

[10] A. Blokland, A.H. Anderson, Effect of low frame-rate video on intelligibility of speech, Speech Commun. 26 (1–2) (1998) 97–103, https://doi.org/10.1016/S0167-6393(98)00053-3.

[11] T. Saitoh, R. Konishi, A study of influence of word lip-reading by change of frame rate, World J. Int. Linguist. Assoc. (2010) 400–407.

[12] W. Mattheyses, L. Latacz, W. Verhelst, Comprehensive many-to-many phoneme-to-viseme mapping and its application for concatenative visual speech synthesis, Speech Commun. 55 (7–8) (2013) 857–876, https://doi.org/10.1016/j.specom.2013.02.005.

[13] D. Jachimski, A. Czyzewski, T. Ciszewski, A comparative study of English viseme recognition methods and algorithms, Multimed. Tools Appl. 77 (13) (2018) 16495–16532, https://doi.org/10.1007/s11042-017-5217-5.

[14] D.S. Alexandre, J.M.R.S. Tavares, Introduction of human perception in visualization, Int. J. Imaging 4 (10A) (2010) 60–70.

[15] S. Miglani, K. Garg, Factors affecting efficiency of K-means algorithm, Int. J. Adv. Res. Technol. 2 (2013) 85–87.

[16] M. Mohajer, K.-H. Englmeier, V.J. Schmid, A Comparison of Gap Statistic Definitions with and Without Logarithm Function, 2011. http://arxiv.org/abs/1103.4767.

[17] H.L. Bear, R.W. Harvey, Y. Lan, Finding Phonemes: Improving Machine Lip-Reading, 2017, pp. 115–120. http://arxiv.org/abs/1710.01142.

[18] B. Aschenberner, C. Weiss, Phoneme-Viseme Mapping for German Video-Realistic Audio-Visual-Speech-Synthesis, Institut Für Kommunikationsforschung Und Phonetik, Universität Bonn, 2005, pp. 1–11.

[19] U. Meier, R. Stiefelhagen, J. Yang, A. Waibel, Towards unrestricted lip reading, Int. J. Pattern Recognit. Artif. Intell. 14 (5) (2000) 571–585, https://doi.org/10.1142/S0218001400000374.

[20] E. Setyati, S. Sumpeno, M.H. Purnomo, K. Mikami, M. Kakimoto, K. Kondo, Phoneme-viseme mapping for Indonesian language based on blend shape animation, IAENG Int. J. Comput. Sci. 42 (3) (2015) 1–12.

[21] S.D. Lalitha, K.K. Thyagharajan, A study on lip localization techniques used for lip reading from a video, Int. J. Appl. Eng. Res. 11 (1) (2016) 611–615.

[22] A.A. Montgomery, P.L. Jackson, Physical characteristics of the lips underlying vowel lipreading performance, J. Acoust. Soc. Am. 73 (6) (1983) 2134–2144, https://doi.org/10.1121/1.389537.

[23] P. Damien, N. Wakim, M. Egéa, Phoneme-viseme mapping for modern, classical arabic language, in: 2009 International Conference on Advances in Computational Tools for Engineering Applications, ACTEA 2009, vol. 2(1), 2009, pp. 547–552, https://doi.org/10.1109/ACTEA.2009.5227875.

[24] T.J. Hazen, K. Saenko, C.H. La, J.R. Glass, A segment-based audio-visual speech recognizer: data collection, development, and initial experiments, in: ICMI'04—Sixth International Conference on Multimodal Interfaces, 2004, pp. 235–242.

[25] J. Melenchón, J. Simó, G. Cobo, E. Martínez, A. La, U.R. Llull, Objective Viseme Extraction and Audiovisual Uncertainty : Estimation Limits between Auditory and Visual Modes, 2007, p. 2007.

[26] A.N. Mishra, M. Chandra, A. Biswas, S.N. Sharan, Hindi phoneme-viseme recognition from continuous speech, Int. J. Signal Syst. Eng. 6 (3) (2013) 164–171, https://doi.org/10.1504/IJSISE.2013.054793.

[27] P. Upadhyaya, O. Farooq, M.R. Abidi, P. Varshney, Comparative study of visual feature for bimodal Hindi speech recognition, Arch. Acoust. 40 (4) (2015) 609–619, https://doi.org/10.1515/aoa-2015-0061.

[28] P. Varshney, O. Farooq, P. Upadhyaya, Hindi viseme recognition using subspace DCT features, Int. J. Appl. Pattern Recognit. 1 (3) (2014) 257, https://doi.org/10.1504/ijapr.2014.065768.

[29] A. Brahme, U. Bhadade, Phoneme visem mapping for Marathi language using linguistic approach, in: Proceedings—International Conference on Global Trends in Signal Processing, Information Computing and Communication, ICGTSPICC 2016, 2017, pp. 152–157, https://doi.org/10.1109/ICGTSPICC.2016.7955288.

[30] K.T. Bibish Kumar, R.K. Sunil Kumar, E.P.A. Sandesh, V.L. Lajish, A comparative study of lip region segmentation in different colour space for lip reading in Indian context, Int. J. Tomogr. Simul. 33 (1) (2020) 73–89.

[31] N. Ahmad, S. Datta, D. Mulvaney, O. Farooq, A comparison of visual features for audiovisual automatic speech recognition, J. Acoust. Soc. Am. 123 (5) (2008) 3939, https://doi.org/10.1121/1.2936016.

[32] D. Stewart, R. Seymour, J. Ming, Comparison of image transform-based features for visual speech recognition in clean and corrupted videos, Eurasip J. Image Video Proc. 2008 (2008) (2008) 1–9, https://doi.org/10.1155/2008/810362.

[33] C. Sui, M. Bennamoun, R. Togneri, Visual speech feature representations: recent advances, in: Advances in Face Detection and Facial Image Analysis, 2016, pp. 377–396.

[34] D. Websdale, B. Milner, Analysing the importance of different visual feature coefficients, in: FAAVSP, 3, 2015, pp. 137–142.

[35] O. Farooq, P. Upadhyaya, O. Farooq, P. Varshney, A. Upadhyaya, Enhancement of VSR Using Low Dimension Visual Feature Enhancement of VSR Using Low Dimension Visual Feature, November, 2013, https://doi.org/10.1109/MSPCT.2013.6782090.

[36] N. Puviarasan, S. Palanivel, Lip reading of hearing impaired persons using HMM, Expert Syst. Appl. 38 (4) (2011) 4477–4481, https://doi.org/10.1016/j.eswa.2010.09.119.

[37] S.S. Morade, Visual Lip reading using 3D-DCT and 3D-DWT and LSDA, Int. J. Comput. Appl. 136 (4) (2016) 7–15.

[38] S.S. Morade, S. Patnaik, Lip reading by using 3-D discrete wavelet transform with Dmey wavelet, Int. J. Image Process. 8 (2014) 384–396.

[39] R. Rajavel, P.S. Sathidevi, Static and dynamic features for improved HMM based visual speech recognition, in: Proceedings of the First International Conference on Intelligent Human Computer Interaction, 2009, pp. 184–194, https://doi.org/10.1007/978-81-8489-203-1_17.

[40] H. Xiaopeng, Y. Hongxun, W. Yuqi, C. Rong, A PCA based visual DCT feature extraction method for lip-reading, in: Proceedings—2006 International Conference on Intelligent Information Hiding and Multimedia Signal Processing, IIH-MSP 2006, December 2006, 2006, pp. 321–324, https://doi.org/10.1109/IIH-MSP.2006.265008.

[41] S. Alizadeh, R. Boostani, V. Asadpour, Lip feature extraction and reduction for hmm-based visual speech recognition systems, in: International Conference on Signal Processing Proceedings, ICSP, 2008, pp. 561–564, https://doi.org/10.1109/ICOSP.2008.4697195.

[42] J. He, H. Zhang, Research on visual speech feature extraction, in: Proceedings—2009 International Conference on Computer Engineering and Technology, ICCET 2009, vol. 2, 2009, pp. 499–502, https://doi.org/10.1109/ICCET.2009.63.

[43] B.D. Baswaraj, A. Govardhan, P. Premchand, Active contours and image segmentation: the current state of the art, Global J. Comp. Sci. Technol. Graph. Vis. 12 (11) (2012).

[44] A. Biswas, P.K. Sahu, A. Bhowmick, M. Chandra, VidTIMIT audio visual phoneme recognition using AAM visual features and human auditory motivated acoustic wavelet features, in: 2015 IEEE 2nd International Conference on Recent Trends in Information Systems, ReTIS 2015—Proceedings, 2004, 2015, pp. 428–433, https://doi.org/10.1109/ReTIS.2015.7232917.

[45] N. Li, N. Lefebvre, R. Lengellé, Kernel hierarchical agglomerative clustering: Comparison of different gap statistics to estimate the number of clusters, in: ICPRAM 2014—Proceedings of the 3rd International Conference on Pattern Recognition Applications and Methods, January, 2014, pp. 255–262, https://doi.org/10.5220/0004828202550262.

[46] T.S. Madhulatha, An overview on clustering methods, IOSR J. Eng. 2 (4) (2012) 719–725. http://arxiv.org/abs/1205.1117.

[47] A.K. Jain, Data clustering: 50 years beyond K-means, Pattern Recogn. Lett. 31 (8) (2010) 651–666, https://doi.org/10.1016/j.patrec.2009.09.011.

[48] R. Tibshirani, G. Walther, T. Hastie, Estimating the number of clusters in a data set via the gap statistic, J. R. Stat. Soc. B 63 (2) (2001) 411–423.

[49] A.D. Gritzman, D.M. Rubin, A. Pantanowitz, Comparison of colour transforms used in lip segmentation algorithms, Signal Image Video Process. 9 (4) (2015) 947–957, https://doi.org/10.1007/s11760-014-0615-x.

[50] M. McLaren, Y. Lei, Improved Speaker Recognition Using DCT Coefficients as Features, 2015, pp. 4430–4434.

[51] K.T. Bibish Kumar, S. John, K.M. Muraleedharan, R.K. Sunil Kumar, Audio-visual asynchrony in Malayalam phonemes and allophones, Int. J. Recent Technol. Eng. 8 (3) (2019) 8359–8362, https://doi.org/10.35940/ijrte.C6468.098319.

CHAPTER 7

Closed-set speaker identification system based on MFCC and PNCC features combination with different fusion strategies

Musab T.S. Al-Kaltakchi[a], Mohammed A.M. Abdullah[b], Wai L. Woo[c], and Satnam S. Dlay[d]

[a]Department of Electrical Engineering, College of Engineering, Mustansiriyah University, Baghdad, Iraq
[b]Computer and Information Engineering Department, College of Electronics Engineering, Ninevah University, Mosul, Iraq
[c]Department of Computer and Information Sciences, Northumberia University, Newcastle upon Tyne, United Kingdom
[d]School of Electrical and Electronic Engineering, Newcastle University, Newcastle upon Tyne, United Kingdom

7.1. Background

Speech is commonly exploited as a human biometric due to the unique characteristics of an individual's voice [1–3]. In speaker identification and verification tasks, choosing the best features to capture this information is one of the most important issues. Mel frequency cepstral coefficient (MFCC) features are widely used for this purpose [3, 4]. However, to improve speaker identification accuracy (SIA), MFCC features are fused with inverse MFCC features (IMFCC) [5]. This approach, however, was limited by the number of Gaussian mixture model (GMM) components. Thus improvement in the recognition rate is still low. In addition [6], we propose combining phase information with MFCC features to improve speaker identification. According to Ref. [7], a text-independent speaker identification system can be achieved with MFCC features and by using a back–propagation neural network (BPNN) for classification. The drawback of this system is its complexity and long training time. Others have proposed the wavelet transform for feature extraction and vector quantization for the modeling technique, but poor identification performance was achieved on the database used (15 speakers) [8]. Moreover, three scenarios for speaker identification were presented by Ref. [9], exploiting the GFM. However,

Applied Speech Processing
https://doi.org/10.1016/B978-0-12-823898-1.00001-1
© 2021 Elsevier Inc.
All rights reserved.
147

the identification rate using the NIST 2003 database was poor. In addition, other researchers, such as Ref. [10], have examined large population speaker identification where the total variability space is used to capture both the speaker and channel variabilities by the I-vector. Their main challenge was providing a suitable database, so 1000 speakers were taken from a non-standard YouTube database. The system gave a good performance for large observation periods (20 s and longer), but it was less efficient for shorter speech lengths (e.g., 10 s). In another study, a large population was achieved by using fuzzy clustering, as in Ref. [11], where hierarchical tree decisions were employed for speaker identification. The authors reported an improvement in the system performance using fusion experiments on 3805 speakers subjected to AWGN; however, the authors have not tested their method using realistic noise factors.

In this chapter, a better identification performance is achieved for large populations compared with other work [5] due to the following points: The power normalized cepstral coefficient (PNCC) features are robust to different types of noise and can sometimes achieve better SIA compared with MFCC and perceptual linear prediction (PLP) features even in a clean environment [12]. Therefore, combining PNCC and MFCC features provides a robust performance for original speech recordings (OSRs) and noisy environments. In contrast with prior work, the proposed system also has the potential to achieve enhanced SIA by removing and reducing sensitivity due to the channel between the speaker and microphone together with handsets by using feature warping (FW) and cepstral mean and variance normalization (CMVN) [5]. Moreover, instead of modeling individual speakers with limited data only by a GMM, as in a previous study, GMM–UBM is used based on modeling strategy as in Ref. [3] utilizing all speakers' data to increase the number of mixtures and thereby enhancing the identification rate. Furthermore, this chapter studies a number of late fusion methods that include weighted sum, maximum, and mean fusion of the combination of the features scores as methods to improve SIA [13, 14]. Also, early fusion and a combination of early and late fusion are used [13]. Moreover, speaker identification with late fusion for static and dynamic features is also included in this work by using the vertical concatenated fusion methods between MFCC and PNCC features. Furthermore, a new method of speaker identification system is accomplished based on the creation of a new score vector from score vectors used in late, early, and late fusion, and the concatenated of static and dynamic features mentioned above. This new vector was achieved via assuming that all fusion score vectors are statistically independent, which are essentially acquired from different feature dimensions 16, 32, and 39. Our contributions in this chapter can be summarized as follows.

- The PNCC and MFCC features are combined to provide robust performance for both original speech and noisy environments.
- Novel fusion strategies are investigated for text-independent speaker identification.
- Four main simulations for SIA are presented for the TIMIT database: feature-based early fusion, score-based late fusion, early-late fusion, and the statistically independent normalized scores fusion for all the previous scores.

The organization for this chapter is as follows: Section 7.2 focuses on biometric speaker identification framework. Section 7.3 reviews fusion strategies. Section 7.4 includes the simulation setup. Section 7.5 presents an overview of the related work. Section 7.6 shows all original speech recording simulation results for this chapter, followed by discussions in Section 7.7. Finally, Section 7.8 presents conclusions.

7.2. Biometric speaker identification framework

7.2.1 Feature extraction and feature normalization

In this work, two feature extraction methods are combined: MFCC and PNCC. A 16-feature dimension was selected from both MFCC and PNCC for each input frame including the average long-power (features at zero-order) C_0 and Pc_0 for both MFCC and PNCCC, respectively [3, 13, 14]. The speech samples were filtered with a preemphasis filter by using a first-order FIR high-pass filter with emphasis coefficient 0.96 [5, 15]. Hamming windowing was exploited for both MFCC and PNCC features with frame duration of 16 milliseconds with 50% interframe rate overlap [5, 16]. For more information about both features, interested readers can refer to Refs. [17–21].

Feature normalization is also adopted by using FW and CMVN for both MFCC and PNCC features. The main purpose of FW is to produce a stronger representation of the distribution for each cepstral feature. For a specified time interval, warping the distribution of a stream of cepstral features to match the normal Gaussian is called feature warping [22]. FW and CMVN approaches are used to improve the SIA for the system as well as reduce sensitivity to the mismatch between types of telephone handsets. They can also help to reduce linear channel effects [23, 24]. The features and feature compensations are developed in Refs. [13, 14].

7.2.2 Acoustic modeling and matching

Modeling a set of speaker classes is one of the most important stages of the recognition task. In GMMs, each speaker can be represented by a finite weighted mixture of multivariate Gaussian components defined by the mean and covariance parameters, as in Eqs. (7.1), (7.2) [5, 25, 26]:

$$p(x|\lambda) = \sum_{i=1}^{M} \omega_i p_i(x) \qquad (7.1)$$

where ω_i is the ith mixture weight, and

$$p_i(x) = \frac{1}{(2\pi)^{\frac{D}{2}}|\Sigma_i|^{\frac{1}{2}}} \exp\left\{-\frac{1}{2}(x-\mu_i)^T \Sigma_i^{-1}(x-\mu_i)\right\} \qquad (7.2)$$

where x is a D-dimensional random feature vector, M is the number of Gaussian mixture components. For each speaker model there is a parameter set $\lambda = \{\omega_i, \mu_i, \Sigma_i\}$, $i = 1, ..., M$, μ_i and Σ_i are, respectively, the mean and covariance parameters of the ith component density and $(\cdot)^T$ denotes the transpose operator [27]. Diagonal covariance matrices are assumed in this work instead of full covariance matrices as in Refs. [3, 25], which is more efficient computationally with no reduction in SIA. In most practical applications of GMMs in speaker identification, only limited training data are available, therefore we use a different approach to learn these models as described next [13, 14].

7.2.2.1 Universal background model and GMM-UBM

Earlier, the GMM was used as a modeling paradigm in speaker recognition systems. However, this method has major drawbacks such as unseen and insufficient data and hence a degradation in speaker recognition performance is noticed, especially when the number of speakers increases. Therefore, a combination of GMM with a universal background model (UBM) was proposed to solve the problems of GMM [3], and this is exploited in this chapter. One of the most important reasons for using UBM is to overcome the problem of insufficient training data as well as unseen data. For understanding UBM, imagine it as a large pool in which all speaker training sets are used together with the expectation-maximization (EM) method. As a consequence of training with a large amount of data, a greater number of parameters is estimated thereby increasing the dimensionality of mixtures to cover all speakers, which will improve system performance. In addition, the individual speaker models are trained by the maximum a posteriori (MAP) adaptation initialized by the UBM with the training data for each particular speaker. This approach, by training on large data for the UBM followed by adapting it for S different speakers, increases the dimensionality of the models. Therefore this coupling between the UBM (large training data) and individual speaker models (small amount of data) makes the GMM-UBM system better able to model unseen data through having estimates with sufficient parameters to increase the mixture dimensionality (number of mixtures), thus improving the identification rate. The UBM represents an effective model for all nonhypothesized speakers (in practice all the training speakers), and is defined as $p\left(X|\,\lambda_{\overline{hyp}}\right)$ where X is the corresponding $T_{\mathcal{F}}$ feature vectors, $X = [x_1, \ldots x_{T_{\mathcal{F}}}]$, and $\lambda_{\overline{hyp}}$ is the corresponding parameters set. This speaker independent model can then be used to improve speaker identification performance [13, 14].

7.2.2.2 Adaptation of speaker models

The parameters of the speaker models are taken from old estimates from the UBM training and the training data of the individual speakers. To control the balance between the new and old estimates, adaptation coefficients are used for weights, means, and variances, represented by α_i^w, α_i^m, and α_i^v, respectively. The adaptation coefficients are used in the following equations [13, 14, 25]:

$$\hat{\omega}_i = \left[\alpha_i^\omega n_i / T_{\mathcal{F}} + \left(1 - \alpha_i^\omega\right)\omega_i\right]\gamma \tag{7.3}$$

$$\hat{\boldsymbol{\mu}}_i = \alpha_i^m E_i(x) + \left(1 - \alpha_i^m\right)\boldsymbol{\mu}_i \tag{7.4}$$

$$\hat{\boldsymbol{\sigma}}_i^{\,2} = \alpha_i^v E_i\left(x^2\right) + \left(1 - \alpha_i^v\right)\left(\sigma_i^2 + \boldsymbol{\mu}_i^2\right) - \hat{\boldsymbol{\mu}}_i^{\,2} \tag{7.5}$$

where γ is the scale factor to assure all adapted mixture weights have a unity summation. As in Ref. [25], the data-dependent mixing coefficients are calculated as:

$$\alpha_i^\rho = \frac{n_i}{n_i + r^\rho} \tag{7.6}$$

$$n_i = \sum\nolimits_{t=1}^{T_{\mathcal{F}}} \Pr\left(i|\,x_t\right) \tag{7.7}$$

$$\Pr\left(i|\,x_t\right) = \frac{\omega_i p_i\left(x_t\right)}{\sum\nolimits_{j=1}^{M} \omega_j p_j\left(x_t\right)} \tag{7.8}$$

$$E_i(x) = \frac{1}{n_i}\sum\nolimits_{t=1}^{T_{\mathcal{F}}} \Pr\left(i|\,x_t\right)x_t \tag{7.9}$$

$$E_i\left(x^2\right) = \frac{1}{n_i}\sum\nolimits_{t=1}^{T_{\mathcal{F}}} \Pr(i|\ x_t)x_t^2 \qquad (7.10)$$

where r^ρ is a fixed relevance factor, i is the mixture in the UBM, $T_{\mathcal{F}}$ is the number of feature vectors, and $Pr(i|x_t)$ is the probabilistic alignment of the training vectors in the UBM mixture components [26, 25]. For the initial UBM training, $final_{iter} = 20$; the MAP adaptation the relevance factor $r^\rho = 10$, $\rho \in \{m, \omega, v\}$; and $N_{mix} \in \{8, 16, 32, 64, 128, 256, 512\}$; $\alpha_i^\rho \in [0, 1]$, where N_{mix} is the number of Gaussian components and $final_{iter}$ is the number of EM iterations. More details of the parameters and how they are used in the adaptation of speaker models can be found in Refs. [14,25]. Depending on the counts of data n_i, if $\alpha_i^\rho \simeq 0$ for a speaker, the estimate relies more on the old sufficient statistics (low probabilistic count), while $\alpha_i^\rho = 1$ relies only on the new trained parameters (high probabilistic count). The relevance factor r^ρ is used as a control between the new and old parameters [25, 26].

7.2.2.3 Maximum log-likelihood scores

Matching between training and testing is carried out by LLR. According to the Bayesian adaptation learning formula to apply the MAP adaptation, maximum log-likelihood should be achieved. The maximum a posteriori probability can be determined using Eq. (7.11) [26]:

$$S = \arg\max_{1\leq k\leq S} P_r(\lambda_k|\ x) = \arg\max_{1\leq k\leq S} \frac{p(x|\ \lambda_k)P_r(\lambda_k)}{p(x)} \qquad (7.11)$$

where S is a set of speakers ($S = \{1, 2, ..., S\}$), which are represented by the GMM models $\lambda_1, \lambda_2, ..., \lambda_S$. The second part of Eq. (7.11) is Bayes' rule. Then, maximum likelihood classification can be derived as in Eq. (7.12) by assuming equally likely speakers with $P_r(\lambda_k) = \frac{1}{S}$ and for all speaker models, $p(x)$ is the same. The maximum likelihood can be determined as in Eq. (7.12) [26]:

$$S = \arg\max_{1\leq k\leq S} p(x|\ \lambda_k) \qquad (7.12)$$

There are two reasons to use the log function in the matching of likelihood. First is that the log function has a monotonically increasing property that makes the maximum position unchanged after taking the log-likelihood for Gaussian models. Second, the log function cancels the exponential function of the GMM. The major purpose of ML estimation is to compute the speaker model parameters that maximize the likelihood of the GMM. For a nonlinearity function such as GMM, direct maximization is not possible.

Therefore estimation ML can be done iteratively by using the EM algorithm. In this technique, initialization is started by choosing the initial model, then an expectation step is performed that probabilistically aligns vectors to generate a new model. The maximization step is achieved by updating model parameters such that they are greater or equal to the initial model and then repeated until convergence is reached [25]. From a test speech signal (unknown speaker), features are extracted that form the inputs to the speaker models (i.e., all speaker models $S = \{\lambda_1, \lambda_2, ..., \lambda_S\}$). The log-likelihood scores are taken for the GMM-UBM system for each trial, which forms a two-dimensional array with a model-test set with a length of 57,600 scores to represent the multiplication between 120 models with 480 tests (4 testing files for each speaker out of 120 speakers). So the trials are represented by model-test sets such as (Model 1, Test 1), ..., (Model 120, Test 1) to describe the scoring between all speaker models against the first test. However, each speaker has four tests. This will produce (Model 1, Test 2) ... (Model 120, Test 2) and so on for 480 tests such as (Model 1, Test 480) to (Model 120, Test 480). The log-likelihood scores are calculated as in Refs. [13, 14, 25].

$$LLR(X) = \log_e p(X| \lambda_{GMM}) - \log_e p(X| \lambda_{UBM}) \qquad (7.13)$$

In this work, four combination vectors of log-likelihood scores are produced based on normalization and feature types. Each log-likelihood score vector has a length of 57,600 scores representing120 speakers by the scoring between 120 training speech files against 480 test files (four speech samples for each speaker). In training and testing the resulting speaker models are scored with a maximum likelihood approach. For the final decision, the maximum log-likelihood approach is used for speaker identification by taking the maximum score for each set of test scores for each speaker model, as in Refs. [3, 26, 28].

7.3. Speaker identification systems with fusion strategies

7.3.1 System 1: Speaker identification system with late fusion

Fig. 7.1 shows the late fusion speaker identification system. Four combinations of MFCC and PNCC features with FW and CMVN are tested in this scenario. Late fusion is represented by score-based fusion and includes three methods: maximum, mean, and weighted sum for 16 feature dimensions. Point A in Fig. 7.1 represents the scores for the normalized PNCC features (FWPNCC or CMVNPNCC). Point B denotes the normalized CMVN

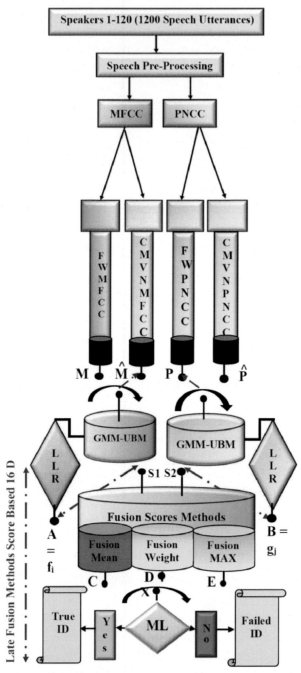

Fig. 7.1 Flowchart of the speaker identification system multibases (16 dimensions) with/without (W/WO) late fusion.

features (FWMFCC or CMVNMFCC). Hence, by connecting either point A or B to point X, four different identification systems can be produced. The first two result from the MFCC features normalized with either FW or CMVN. Similarly, the other two stem from the PNCC features normalized with FW or CMVN. To develop late fusion, either point A or B can be connected to point S1 to produce maximum, mean, and weighted sum fusion [13, 14].

Three late fusion approaches are adopted: depending on the features and normalization methods, four combinations of log-likelihood scores are constructed. These are: $f_1 = $ FW MFCC scores vector (FWMFCC), $f_2 = $ CMVN MFCC scores vector, $g_1 = $ FW PNCC scores vector (FWPNCC), and $g_2 = $ CMVN PNCC scores vectors. These score vectors are found before the fusion process and form the following composite vectors as (7.14), (7.15).

$$f_i = \begin{bmatrix} f_1 \\ f_2 \end{bmatrix} \qquad (7.14)$$

$$g_j = \begin{bmatrix} g_1 \\ g_2 \end{bmatrix} \qquad (7.15)$$

To produce four types of maximum fusion scores vectors of $fmax_{11}$, $fmax_{12}$, $fmax_{21}$, and $fmax_{22}$, the row-wise maximum is evaluated as in Eq. (7.16).

$$fmax_{ij} = \max\left(f_i, g_j\right) \qquad (7.16)$$

where $fmax_{ij}$ is the fusion maximum scores vector, $i, j = 1, 2$.

Similarly, to produce the four mean fusion score vectors of $fmean_{11}$, $fmean_{12}$, $fmean_{21}$, and $fmean_{22}$, Eq. (7.17) is used to calculate the fusion mean.

$$fmean_{ij} = \left(f_i + g_j\right)/2 \qquad (7.17)$$

where $fmean_{ij}$ is the fusion mean scores vector, $i, j = 1, 2$.

In addition, a linear weighted sum fusion of scores is used for the score vectors $fweight_{11}$, $fweight_{12}$, $fweight_{21}$, and $fweight_{22}$, as in Eq. (7.18).

$$fweight_{ij} = \omega_\beta f_i + \left(1 - \omega_\beta\right)g_j \qquad (7.18)$$

where $\beta = 1, 2, 3, 4$ and $\omega_1, \omega_2, \omega_3,$ and $\omega_4 = 0.9, 0.8, 0.77,$ and 0.7, respectively. Both i and j take values of 1 and 2, therefore $fweight_{ij}$ takes one of four values $fweight_{11}$, $fweight_{12}$, $fweight_{13}$, and $fweight_{22}$, and *fweight*$_{11}$ is the linear combination of f_1 and g_1. Likewise, *fweight*$_{12}$ is the linear combination of f_1 and g_2 and so on. For each *fweight*$_{ij}$, ω_β can take on one of three values ($\omega_\beta \in \{0.7,0.9,0.8\}$), which are empirically chosen to give the best SIA. This is because these values represent the MFCC coefficients, and assigning more weight to the MFCC coefficients improves performance with clean speech.

7.3.2 System 2: Speaker identification system with early fusion and early-late fusion

In Fig. 7.2 four systems can be obtained from the vertical concatenation of normalized MFCC and PNCC features. Since each one has 16 feature dimensions, the resultant early fusion vector has 32 feature dimensions. This system can be further improved using the late fusion technique [13]. This process is called early fusion because the fusion is performed before modeling the system using GMM–UBM. The system can be described as [13]: $M = \{M_1, M_2, ..., M_k\}$, $\hat{M} = \{\hat{M}_1, \hat{M}_2, ..., \hat{M}_k\}$, $P = \{P_1, P_2, ..., P_k\}$, $\hat{P} = \{\hat{P}_1, \hat{P}_2, ..., \hat{P}_k\}$, where $\hat{M} =$ CMVNMFCC feature matrices, $M =$ FWMFCC feature matrices, $P =$ FWPNCC feature matrices, $\hat{P} =$ CMVNPNCC feature matrices, and k is equal to 480 in the testing phase and 720 in the training phase. Each feature matrix has the dimension of $D \times N$, where D is the feature dimension and is equal to 16 while N is the number of frames and is equal to 1000. Early feature fusion can be represented by Eq. (7.19):

$$H_{i,j} = [MFCC_i \, PNCC_j], \quad i,j = 1,2. \tag{7.19}$$

where $MFCC_i$ is normalized MFCC feature matrices before modeling that contain either $MFCC_1$ as M, or $MFCC_2$ as $\hat{M}$. Likewise, $PNCC_i =$ is the normalized PNCC feature matrices before modeling where $PNCC_1$ is P and $PNCC_2$ is $\hat{P}$. In the same manner, the system will produce four lines of early fused features by connecting either point A or B to the maximum likelihood (ML) point (X). In addition, late fusion can be integrated with the system to produce early-late fusion by attaching points A and B to points S1 and S2.

7.3.3 System 3: Speaker identification system with late fusion for concatenated static and dynamic features

This system presented here is similar to system 1, but the main difference is the increase in feature dimensions to include the static and dynamic MFCC

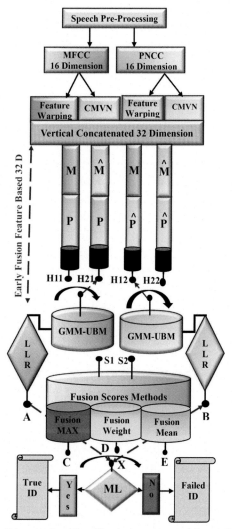

Fig. 7.2 Flowchart of the speaker identification system multibases (32 dimension) early fusion with/without late fusion.

and PNCC features to 39 features per frame. This is achieved by concatenating 13 MFCC/PNCC features (original features) with temporal derivatives including 13 features from the first-order derivative (Delta) as well as the corresponding 13 features from the second-order derivatives (Delta-Delta) to yield 39 features such as $FeatureDim(39MFCC) = 13MFCC + 13\Delta MFCC + 13\Delta\Delta MFCC$; likewise for PNCC features $FeatureDim(39PNCC) = 13MFCC + 13\Delta PNCC + 13\Delta\Delta PNCC$ [3].

Late fusion methods score based are applied to the normalized methods for MFCC (FWMFCC and CMVNMFCC) with the corresponding scores for PNCC normalized features (FWPNCC and CMVNPNCC). Eq. (7.20) is applied to calculate the Delta feature (first-order derivative) for both MFCC and PNCC [3].

$$d_t = \sum_{\delta=1}^{del} \frac{\delta(c_{t+\delta} - c_{t-\delta})}{2\sum_{\delta=1}^{del} \delta^2} \tag{7.20}$$

where *del* is typically 2, $c_t = [c_0, c_1, ..., c_L]$, $L = 12$, c is either MFCC or PNCC features (13 coefficients) (c_0 + L), d_t is the feature vector for the first-order derivative, and t is the frame time index. Similarly, the acceleration parameter a_t vectors, which represent the second–order derivative vectors, can be produced by replacing c_t with d_t in Eq. (7.17). Therefore in order to boost the original MFCC and PNCC features, temporal derivatives d_t and a_t as dynamic features are concatenated with static features c_t.

This strategy was adopted for MFCC only in speech recognition as well in recent work in speaker recognition [3], as explained in Eq. (7.21) and illustrated in Fig. 7.3. In this chapter, we exploit this technique not only for MFCC features but also for PNCC features.

$$cf_t = \left[c_t^T \ d_t^T \ a_t^T \right]^T \tag{7.21}$$

where cf_t is the concatenated static and dynamic temporal features for MFCC/PNCC with 39 dimensions.

7.3.4 System 4: Speaker identification system with late fusion for normalized independent scores for systems 1, 2, and 3

This system is deduced from the vector scores resulting from the three systems previously discusses. These score vectors are normalized for the constancy of scores by subtracting each vector from the mean for that vector. Then, the result is divided by the standard deviation for this vector. These scores are assumed statistically independent where the dimension for each score vector is 57,600, while these scores are coming from different feature

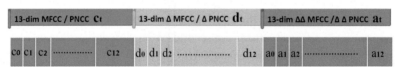

Fig. 7.3 Concatenated MFCC/PNCC static features with the dynamic features with 39 dimension [3].

dimensions (16, 32, and 39). The main purpose is to accomplish new score vectors, which are created via multiplying three independent score vectors developed from different feature dimensions (16, 32, and 39). This new system is quite similar to those in genetic development systems where the new system scores originally come from mixing three different feature dimensions. The score vectors for system 1, system 2, and system 3 are normalized. The equations in (7.22)–(7.24) are adopted to present system 4.

$$sysA_i = \frac{sysa_i - \mu(sysa_i)}{\sigma(sysa_i)} \tag{7.22}$$

$$sysB_i = \frac{sysb_i - \mu(sysb_i)}{\sigma(sysb_i)} \tag{7.23}$$

$$sysC_i = \frac{sysc_i - \mu(sysc_i)}{\sigma(sysc_i)} \tag{7.24}$$

where $sysa_i$, $sysb_i$, and $sysc_i$ represent the four score vectors for system 1, system 2, and system 3, respectively. $sysA_i$, $sysB_i$, and $sysC_i$ are the normalized score vectors for previous systems, respectively. $i = 1, 2, 3, 4$. $sysA_i$ are the normalized score vectors used in system 1 as f_1, f_2, g_1, g_2 for features with 16 dimensions. $sysB_i$ are the normalized score vectors used in system 2 as h_1, h_2, h_3, h_4 for features with 32 dimensions. Finally $sysC_i$ are the normalized score vectors used in system 3 as $\ddot{f}_1, \ddot{f}_2, \ddot{g}_1, \ddot{g}_2$ for features with 39 dimensions. In the new system, it is assumed that all the normalized scores for systems 1, 2, and 3 ($sysA_i$, $sysB_i$, and $sysC_i$) are statistically independent. Multiplications for these score vectors are applied as in Eq. (7.25).

$$sind_i = sysA_i \cdot sysB_i \cdot sysC_i \tag{7.25}$$

where $(\cdot)$ represent the element-wise multiplication, $i = 1, 2, 3,$ and 4.

$$sind_1 = f_1 \cdot h_1 \cdot \ddot{f}_1$$
$$sind_2 = f_2 \cdot h_2 \cdot \ddot{f}_2$$
$$sind_3 = g_1 \cdot h_3 \cdot \ddot{g}_1$$
$$sind_4 = g_2 \cdot h_4 \cdot \ddot{g}_2$$

Four score vectors are produced as a consequence of applying Eq. (7.25), and these vectors are normalized using Eq. (7.26), as explained in Fig. 7.4.

$$sind(norm)_i = \frac{sind_i - \mu(sind_i)}{\sigma(sind_i)} \tag{7.26}$$

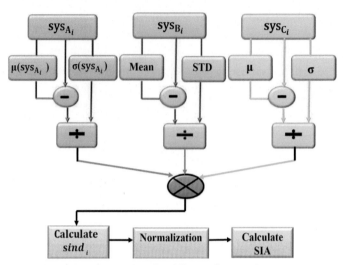

Fig. 7.4 System 4 independent scores for different feature dimensions.

7.4. Simulations setup

In all the simulations, training and testing were conducted on a personal computer with Intel(R) Core(TM) i5-3470 CPU 3.20 GHz, 16 GB of RAM, and Windows 7 as an operating system. The speech sampling frequency used was 16 kHz. We utilized the Texas Instruments/Massachusetts Institute of Technology (TIMIT) database in all of our experiments, as this database is widely exploited in different work [3, 13, 14].

There are eight dialects in the TIMIT database. In this chapter both Dialect Region 1 (DR1) from New England and DR4 from South Midland are employed. The main parameters used in the experiments are explained in Table 7.1. These parameters can be categorized according to the database, DR, sampling rate, window size and frameshift, preemphasis factor, window type, number of speakers used, number of samples per speaker, the total number of samples used, training and testing samples per speaker, and finally, the average duration for each speech sample [13, 14].

7.5. Comparisons with related work

Table 7.2 presents a comparison among all aspects between the work in Ref. [5] and all other simulations in this chapter. In addition, Table 7.2 gives a summary of all simulation infrastructures for this chapter compared with those used in Ref. [5], such as speaker's DR, features and feature dimensions

Table 7.1 Experimental parameters for the work in Ref. [5] and in all proposed simulations in this chapter.

Database	TIMIT
Dialect	DR1 and DR4
Sampling frequency	16,000
Window size	16 ms
Frame shift	8 ms
Preemphasis factor	0.96
Window type	Hamming
Number of speakers	120
No. of samples per speaker	10
Total samples used	1200
Training samples	6 per speaker (total 720)
Testing samples	4 per speaker (total 480)
Average sample duration	8 s
System environment	Original speech recording

Table 7.2 Main comparison between the work in Ref. [5] and the proposed algorithm.

Aspects	Methods in Ref. [5]	Proposed simulations
Speakers DR	DR1 & DR4	49 DR1& 71 DR4
Features	1-MEL 2-IMEL	1-MFCC 2-PNCC
Features dim	N/A	16
Feature norm	Not used	FW and CMVN
Modeling	GMM	GMM-UBM
GMCs (mixtures)	{8, 16, 32}	{8, 16, 32, 64, 128, 256, 512}
Classifier	LLR	LLR
Fusion types	Fusion Weight	Early, late, early-late, concatenated statistic d score independent fusion
SIA	93.88%	**95%**
System environment	Clean	Clean and noisy

used, feature compensation methods (normalization), modeling, Gaussian mixture components (GMCs), type of classifier, fusion types, speaker identification accuracy, and the system environment. The work in Ref. [5] used 120 randomly chosen speakers from two dialect regions, DR1 and DR4. Of the speakers, 49 were taken from DR1 and 71 were taken from DR4. Furthermore, in Ref. [5] MEL was used without employing any type of normalization. In this work, two types of features are employed. The first are

Table 7.3 SIA results for original speech recordings as in Ref. [5].

SIA results for work in Ref. [5]			
Methods	Mix8	Mix16	Mix32
MEL	67.35%	74.36%	71.43%
IMEL	55.10%	58.97%	53.06%
Fused $\omega_1 = 0.5$	79.59%	87.75%	83.67%
Fused $\omega_2 = 0.7$	88.2%	90.31%	89.15%
Fused $\omega_3 = 0.77$	89.8%	**93.88%**	91.84%
Fused $\omega_4 = 0.8$	89.8%	91.84%	91.84%

robust for noise (PNCC) and the second are efficient for original speech recordings (MFCC). In addition, feature normalization is investigated to solve linear channel effect problems. Table 7.3 shows results for the approach in Ref. [5] with three different numbers of GMCs {8, 16, 32}. The highest SIA achieved in Table 7.3 is 93.88% at a mixture size of 16. Continuously increasing the number of mixture components decreases SIA, due to limited GMM data training. This is one of the most important problems that we tackle in this work. Furthermore, in Ref. [5], only Mel features are used and fused with Inverse MEL features (IMEL) using fusion weights of 0.5, 0.7, 0.77, and 0.8.

7.6. Simulation results

7.6.1 Simulation results for system 1

Table 7.4 presents the simulation results for system 1, which include SIAs based on late fusion. This table shows the SIA of four combinations of features based on MFCC and PNCC features for different GMCs, namely, 8, 16, 32, 64, 128, 256, and 512 features. These combinations are: FWMFCC, CMVNMFCC, FWPNCC, and CMVNPNCC. It is clear from Table 7.4 that the SIA for the MFCC features is greater as compared with the corresponding results for PNCC features. This is because the MFCC features have a better performance compared with PNCC features in clean environments. The scores for the best SIA for the MFCC features (FWMFCC and FWMCC, i.e., (f_1) or (f_2)) are fused with the corresponding scores for the best SIA for the PNCC features (FWPNCC and FWPNCC, i.e., g_1 or g_2). Then three late fusion methods are applied to the score vectors belonging to the fusion decision; the late fusion methods are weighted sum, maximum,

Table 7.4 Simulation 1: Speaker identification system with late fusion.

Methods	Mix8	Mix16	Mix32	Mix64	Mix128	Mix256	Mix512
		Speaker identification accuracy (SIA %) for different GMCs						
FWMFCC (f_1)	80%	84.17%	89.17%	93.33%	93.33%	93.33%	94.17%
CMVNMFCC (f_2)	77.5%	80.83%	86.67%	91.67%	91.67%	92.5%	90.83%
FWPNCC (g_1)	60%	71.67%	80.83%	86.67%	88.33%	90%	90%
CMVNPNCC (g_2)	70%	74.17%	83.33%	86.67%	90%	89.17%	90.83%
Fusion decision	(f_1,g_2)	(f_1,g_2)	(f_1,g_2)	(f_1,g_2)	(f_1,g_2)	(f_1,g_1)	(f_1,g_2)
Fused $\omega_1 = 0.9$	79.17%	84.17%	90%	91.67%	93.33%	93.33%	**95%**
Fused $\omega_2 = 0.8$	80.83%	83.33%	90%	92.5%	93.33%	94.17%	94.17%
Fused $\omega_3 = 0.77$	80.83%	83.33%	90%	92.5%	93.33%	94.17%	94.17%
Fused $\omega_4 = 0.7$	79.17%	82.5%	89.17%	92.5%	93.33%	94.17%	93.33%
Fusion max	77.5%	75%	85%	90%	94.17%	93.33%	93.33%
Fusion mean	78.33%	80.83%	90%	92.5%	93.33%	94.17%	94.17%

and mean fusion methods. This work presented four main weights; 0.9, 0.8, 0.77, and 0.7. The 0.77 weight is selected to mirror the work in Ref. [5] for comparison purposes. The first highest SIA (95%) is achieved at fusion weight of 0.9 at a mixture size of 512. The second-highest SIA (94.17%) is accomplished at the weighted sums of 0.8, 0.77, and 0.7 as well as the mean fusion at mixture sizes of 512 and 256. However, related to all mixture sizes, the weighted sum fusion appears to be the best fusion method compared with all late fusion methods. There is an increase in SIA by fusing both MFCC and PNCC features together as compared to using each feature alone. In addition, increasing GMC size causes an additional increment in the data trained by GMM–UBM, which increases the accuracy for SIA [13].

7.6.2 Simulation results for system 2

Table 7.5 presents the simulation results for system 2, which uses early fusion and early-late fusion. This table provides four early fusions by vertically concatenating 16 MFCC features (FWMFCC and CMVNMFCC) with 16 corresponding PNCC features (FWPNCC and CMVNPNCC) to produce four combinations each with 32 dimension features. The early fused matrices with 32 feature dimensions before modeling are: $H_{1,1}$, $H_{2,1}$, $H_{1,2}$, and $H_{2,2}$. Early fusion score vectors for the corresponding previous matrices are: h_1, h_2, h_3, and h_4. The greatest SIA (91.67%) for early fusion is

Table 7.5 Simulation 2: speaker identification system with early fusion and early-late fusion.

Methods	Speaker identification accuracy (SIA %) for different GMCs						
	Mix8	Mix16	Mix32	Mix64	Mix128	Mix256	Mix512
h_1 (scores of $H_{1,1}$)	61.67%	66.67%	76.67%	86.67%	90.83%	89.17%	86.67%
h_2 (scores of $H_{2,1}$)	65%	65%	73.33	84.17%	86.67%	88.33%	85.83%
h_3 (scores of $H_{1,2}$)	64.17%	70.83%	75%	85%	90%	91.67%	86.67%
h_4 (scores of $H_{2,2}$)	65%	65%	79.17%	85.83%	91.67%	90%	89.17%
Fusion decision	$h_3 - h_4$	$h_3 - h_4$	$h_1 - h_4$	$h_1 - h_4$	$h_1 - h_4$	$h_3 - h_4$	$h_3 - h_4$
Fused $\omega_1 = 0.9$	64.17%	70%	76.67%	87.5%	90.83%	**92.5%**	86.67
Fused $\omega_2 = 0.8$	65.83%	71.67%	77.5%	88.33%	90.83%	**92.5%**	87.5
Fused $\omega_3 = 0.77$	65.83%	71.67%	77.5%	88.33%	90.83%	**92.5%**	87.5%
Fused $\omega_4 = 0.7$	65%	71.67%	78.33%	89.17	91.67	**92.5%**	88.33%
Fusion max	64.17%	66.67%	80%	88.33%	90	90.83%	88.33%
Fusion mean	65%	68.33%	79.17%	90.83%	90.83%	**92.5%**	89.17%

accomplished at score vector h_3 at a mixture size of 256. Late fusion strategies to the early score vectors are used to improve the SIA by effectively fusing the scores vector for highest SIA between h_1 and h_3 with the corresponding highest SIA for the scores vector between both h_2 and h_4. The greatest SIA accuracy (92.5%) is achieved via fusing h_3 with h_4 at a mixture size of 256 at weighted sum fusion for all weights used and at fusion mean. However, the fusion mean appears to have better performance accuracy than other fusion methods. There is an increase in SIA when fusing and increasing the mixture, whereas there is a decrease in SIA when increasing the size of GMCs to 512. Therefore, the best SIA is achieved at a mixture size of 256.

7.6.3 Simulation results for system 3

Table 7.6 provides the simulation results for system 3, which contains a speaker identification system with late fusion for the concatenation of static and dynamic features. Generally, this table is similar to Table 7.4. The only

Table 7.6 Simulation 3: speaker identification system with late fusion for concatenated static and dynamic features.

Methods	Speaker identification accuracy (SIA %) for different GMCs						
	Mix8	Mix16	Mix32	Mix64	Mix128	Mix256	Mix512
cf of FWMFCC ($\ddot{f}_1$)	40.83%	44.17%	49.17%	66.67%	74.17%	82.5%	80%
cf of CMVNMFCC ($\ddot{f}_2$)	42.5%	47.5%	57.7%	69.17%	84.17%	85%	85.83%
cf of FWPNCC ($\ddot{g}_1$)	36.67%	40%	43.33%	60.83%	67.5%	71.67%	78.33%
cf of CMVNPNCC ($\ddot{g}_2$)	38.33%	40%	48.33%	65%	80%	75%	80%
Fusion decision	($\ddot{f}_2 - \ddot{g}_2$)	($\ddot{f}_2 - \ddot{g}_2$)	($\ddot{f}_2 - \ddot{g}_2$)	($\ddot{f}_2 - \ddot{g}_2$)	($\ddot{f}_2 - \ddot{g}_2$)	($\ddot{f}_2 - \ddot{g}_2$)	($\ddot{f}_2 - \ddot{g}_2$)
Fused $\omega_1 = 0.9$	45.83%	48.33%	61.67%	71.67%	85%	87.5%	**87.5%**
Fused $\omega_2 = 0.8$	45%	50%	63.33%	72.5%	86.67%	86.67%	**87.5%**
Fused $\omega_3 = 0.77$	45%	51.67%	63.33%	72.5%	85%	86.67%	86.67%
Fused $\omega_4 = 0.7$	45%	50.83%	63.33%	73.33%	85%	86.67%	86.67%
Fusion_max	41.67%	43.33%	53.33%	65%	79.17%	82.5%	85%
Fusion mean	47.5%	50.83%	58.33%	72.5%	84.17%	86.67%	**87.5%**

difference is the number of feature dimensions; Table 7.6 uses 39 features, whereas Table 7.4 uses 16. The concatenations are adopted for 13 MFCC features and 39 Delta and Delta-Delta features; likewise for PNCC features. Then fusion is performed between the highest SIA for the concatenated MFCC features $(\ddot{f}_1, \ddot{f}_2)$ with the corresponding PNCC features $(\ddot{g}_1, \ddot{g}_2)$. It is evident from Table 7.6 that the mixture size and the dynamic features represented by the first and second derivatives play a significant role in calculating SIA. Moreover, there is a 50% decrease in SIA at a mixture size of 8 as shown in Tables 7.4 and 7.5, whereas an acceptable SIA can be achieved by increasing the GMCs (128–512). As shown in Table 7.6, maximum SIA of 87.5% occurs at a mixture size of 512 at weights 0.9 and 0.8 as well as at fusion mean when fusing $\ddot{f}_2$ with $\ddot{g}_2$.

7.6.4 Simulation results for system 4

Table 7.7 presents the simulation results for system 4, which contains the fusion for normalized independent scores for systems 1, 2, and 3. Similarly, Tables 7.4–7.6 are constructed. Each score vector in $sind(norm)_i$ is developed from the normalized to the element multiplications for statistically independent vectors in three previous simulations. The highest SIA (93.33%) is achieved at a mixture size of 128 at fusion maximum between $sind(norm)_2$ and $sind(norm)_4$. Generally, the results shown in this table are better than those presented in Tables 7.5 and 7.6. The most important issue is that

Table 7.7 Simulation 4: speaker identification system with late fusion for normalized independent scores for systems 1, 2, and 3.

	Speaker identification accuracy (SIA %) for different GMCs						
Methods	**Mix8**	**Mix16**	**Mix32**	**Mix64**	**Mix128**	**Mix256**	**Mix512**
$sind(norm)_1 = 1$	79.17%	85%	84.17%	89.17%	91.67%	90.83%	92.5%
$sind(norm)_2 = 2$	78.33%	83.33%	86.67%	90%	91.67%	91.67%	90.83%
$sind(norm)_3 = 3$	68.33%	70.83%	76.67%	83.33%	85.83%	85.83%	86.67%
$sind(norm)_4 = 4$	70.83%	71.67%	79.17%	85.83%	88.33%	87.5%	89.17%
Fusion decision	(1,4)	(1,4)	(2,4)	(2,4)	(2,4)	(2,4)	(1,4)
Fused $\omega_1 = 0.9$	78.33%	83.33%	86.67%	89.17%	91.67%	91.67%	90.83%
Fused $\omega_2 = 0.8$	79.17%	83.33%	85.83%	89.17%	91.67%	91.67%	90.83%
Fused $\omega_3 = 0.77$	79.17%	81.67%	85.83%	89.17%	91.67%	91.67%	90%
Fused $\omega_4 = 0.7$	79.17%	80.83%	85.83%	89.17%	91.67%	91.67%	90%
Fusion max	77.5%	78.33%	84.17%	90.83%	**93.33%**	92.5%	92.5%
Fusion mean	77.5%	78.33%	85%	89.17%	90.83%	91.67%	90.83%

the scores used are created from different scores; from three different feature dimensions of 16, 32, and 39. Therefore, similar to genetic development, performance improves by fusion scores with different feature dimensions.

7.7. Discussions

Simulation 1 and Simulation 2 describe early, late, and early-late fusion based on features, scores, and their combination. Fig. 7.5 illustrates the SIA of the cases discussed in Simulation 1. Fig. 7.6 shows the relationship between SIA and different mixture size of GMCs for Simulation 1.

In Fig. 7.5, seven SIA values are represented for each box plot for different dimensions of GMCs {8, 16, 32, 64, 128, 256, and 512}. The first four boxes, which are *red, pink, green,* and *blue,* are utilized to perform SIA in the situation of multibase boxes for different features before using the fusion methods of FWMFCC, CMVNMFCC, FWPNCC, and CMVNPNCC (f_1, f_2, g_1, g_2). However, five boxes display late fusion techniques: fusion weights at $\omega_\beta = 0.9, 0.8$, and 0.7, respectively, which empirically have been found to produce the best SIA as well as mean and maximum fusion. This fusion can be achieved by fusing the highest SIA features between (f_1, f_2), which is f_1 represented by a *red box.* In addition, the highest SIA between

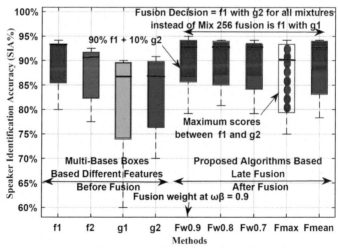

Fig. 7.5 Box plot representing performance in terms of SIA for OSR with multibases boxes with late fusion for 16D relating to Simulation 1.

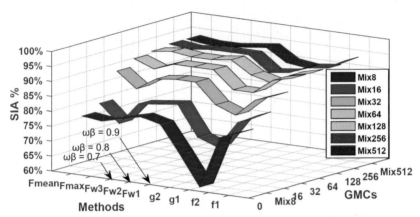

Fig. 7.6 Ribbon plot for OSRs depending on late fusion approaches for 16 dimensions, which relates to Simulation 1.

(g_1, g_2) is g_2, which is represented by the *blue box* for all mixture sizes excluding the mixture size of 256, which is represented g_1.

In order to achieve the desired result, fusion boxes achieved higher SIA against bases boxes. The highest SIA performance is attained (95%) at weighted sum fusion with $\omega_\beta = 0.9$ and a mixture size of 512. It is clear from Fig. 7.5 that late fusion results in greater SIA than PNCC or MFCC features alone, as shown in the multibases boxes.

Fig. 7.6 illustrates how much increasing the GMCs affects SIA performance, illustrated in the form of a 3D ribbon plot. It is obvious that increasing the GMCs from 8 to 512 increases SIA. This is due to the combination of benefits of GMM-UBM together with fusion methods. Even though the best SIA result at GMC 256 is slightly less (94.17%) than the SIA at GMC 512 (95%), the other SIA results for GMC 256 are better than those in mixture size of 512 [13].

Fig. 7.7 describes the second OSR simulation results where the feature dimension is doubled and achieved by concatenating different features. In contrast with Simulation 1, 32 feature dimensions are used instead of 16. According to Fig. 7.7, there are four bases lines (BL1, BL2, BL3, and BL4), which represent the SIA for early fusion. BLN1 is concatenated with 16 features from FWPNCC (P) with the corresponding FWMFCC features (M) to yield $H_{1,1}$ with 32 features. Likewise, BLN2 concatenates the 16 FWPNCC features (P) with the corresponding CMVNMFCC features ($\hat{M}$) to produce $H_{2,1}$ with 32 features. BLN3 concatenates the 16 CMVNPNCC ($\hat{P}$) features with the corresponding FWMFCC features

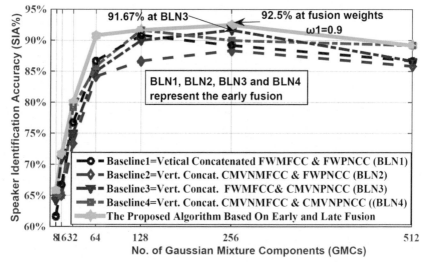

Fig. 7.7 Comparison of SIA in OSR depending on early fusion (multibases lines) against early-late fusion relating to Simulation 2.

(M) to yield $H_{1,2}$ with 32 features. Finally in BLN4, the concatenation of 16 CMVNPNCC ($\hat{P}$) features with the corresponding CMVNMFCC features ($\hat{M}$) produces $H_{2,2}$, yielding 32 features, as in Eq. (7.16).

In Simulation 2, the second part is accomplished by adding late fusion techniques. The scores that have higher SIA between ($H_{2,1}$, $H_{2,2}$) are selected for each GMC and fused by late fusion methods with corresponding scores that have the highest SIA between ($H_{1,1}$, $H_{1,2}$). In the end, the highest SIA from all early-late fusion techniques is taken for each mixture to produce the proposed early-late fusion algorithm based on score-feature combination, shown in Fig. 7.7 as a *green line*. In addition, an early-late fusion algorithm attains slightly higher SIA (92.5%) at GMC 256 compared with the highest SIA (91.67%) at BLN3 on GMC 256, as shown in Fig. 7.7 [13]. According to Fig. 7.8, the bar chart explains the comparison between SIA for late fusion in Simulation 1 with a combination of early-late fusion in Simulation 2. The bar chart provides an indication that late fusion with 256 feature dimensions is better (94.17%) than early and late fusion (92.5%).

Simulation 2 and Simulation 1 are presented with three methods of fusion based on 16 and 32 feature dimensions (FDs). The late fusion scheme with 16 FDs produced the best SIA; the early-late method with 32 FDs produced the next best SIA; and early fusion with 32FDs produced the lowest SIA [13, 14]. Each fusion technique gives a higher SIA than using separate

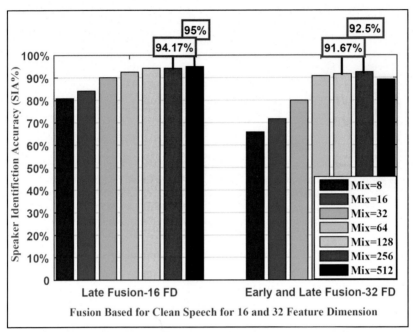

Fig. 7.8 Comparison of SIA performance for OSRs for late fusion (Simulation 1) and early-late fusion (Simulation 2).

normalized PNCC or MFCC features only. It is clear that Simulation 1 (*red curve* in Fig. 7.9) gives the best performance (95%) at a mixture size of 512, while Simulation 4 gives the second-best SIA (93.33%) at a mixture size of 128 (*green curve* in Fig. 7.9). Simulation 2 gives a 92.5% SIA at a mixture size of 256 (*pink curve* in Fig. 7.9), whereas Simulation 3 produces the lowest performance accuracy with SIA of 87.5% at mixture sizes of 256 and 512 (*blue curve* in Fig. 7.9).

It is clear from Simulation curves 1 and 4 in Fig. 7.9 that SIA gradually increases from mixture size 8 to 64, whereas SIA seems to be semistable for the remaining period from mixture size 128 to 512. However, the SIA in Simulations 2 and 3 dramatically increase for mixture sizes 8–64, whereas SIA in Simulation 3 increases up to mixture size of 128 and then tends to be stable to the end at a mixture size of 512. In contrast, SIA in Simulation 2 curve slightly increases for mixtures with sizes spanning from 64 to 256, whereas SIA tends to decrease the endpoint at a mixture size of 512. Furthermore, Fig. 7.9 shows the concatenation between the static and dynamic features for the first- and second-order derivatives in Simulation 3 (*blue dot*

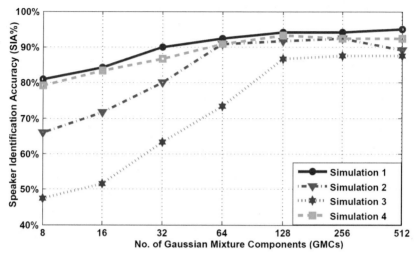

Fig. 7.9 SIA against GMCs for all OSRs.

curve), which gives significant results at mixture sizes 128–512. However, this simulation results in the worst SIA compared with other simulations. In addition, in a noisy environment, the dynamic features are derivatives in the time domain and represented as a multiplication when converted to the frequency domain. This increases noise when increasing frequency and reduces the performance accuracy for speaker identification. It can be deduced from all results that a mixture size of 256 represents the best mixture, which gives the highest SIA.

7.8. Conclusions

In this chapter, four main simulations with fixed-length original speech recordings (129,250 samples, each 8 s in length) were performed to calculate SIA for different GMCs and feature dimensions based on different fusion techniques. These fusion methods included score-based fusion 16 FD (late fusion), feature-based 32 FD (early fusion) and feature score-based 32 FD (early–late fusion), concatenated static and dynamic features (feature-based 39FD), and finally the multiplication of scores independent for different feature dimensions (16, 32, and 39).

Results indicate that late fusion is dominant compared with other fusion methods for evaluating SIA. In addition, late fusion consists of three main fusion methods: weighted sum, mean, and maximum. The Gaussian mixture with a size of 256 gives the highest or second-highest SIA for all OSRs used in this chapter. The highest SIA is 95%, which is achieved at a mixture size of 512 in Simulation 1 with late fusion. We used 16, 32, and 39 feature dimensions, and it is clear from simulations that the best performance accuracy is accomplished with 16 feature dimensions. Fusion-based methods for MFCC and PNCC features result in better accuracy than using each feature alone. Late fusion obtains the first highest SIA of 95%, while fusion by multiplicating different scores obtains the second-best SIA of 93.33%. The combination of early–late fusion results in the next greatest SIA of 92.5%. Early fusion only produces a SIA of 91.67%, whereas late fusion for concatenated static and dynamic features produces the lowest SIA of 87.5%.

The experiments were conducted using clean speech recordings. In the future, to make the work more realistic, we will employ noisy speech recordings.

References

[1] A.K. Jain, A. Ross, S. Prabhakar, An introduction to biometric recognition, IEEE Trans. Circuits Syst. Video Technol. 14 (1) (2004) 4–20.
[2] N. Dey (Ed.), Intelligent Speech Signal Processing, Academic Press, 2019.
[3] S. Sen, A. Dutta, N. Dey, Audio Processing and Speech Recognition: Concepts, Techniques and Research Overviews, Springer, 2019.
[4] R. Togneri, D. Pullella, An overview of speaker identification: accuracy and robustness issues, IEEE Circuits Syst. Mag. 11 (2) (2011) 23–61.
[5] J. Benesty, Springer Handbook of Speech Processing, Springer Science & Business Media, 2008.
[6] R.S.S. Kumari, S.S. Nidhyananthan, G. Anand, Fused MEL feature sets based text-independent speaker identification using Gaussian mixture model, Procedia Eng. 30 (2012) 319–326.
[7] S. Nakagawa, L. Wang, S. Ohtsuka, Speaker identification and verification by combining MFCC and phase information, IEEE Trans. Audio Speech Lang. Process. 20 (4) (2012) 1085–1095.
[8] S. Nandyal, S.S. Wali, S.M. Hatture, MFCC based text-dependent speaker identification using BPNN, Int. J. Signal Proces. Syst. 3 (1) (2015) 30–34.
[9] S.S. Yadav, D.G. Bhalke, Speaker identification system using wavelet transform and VQ modeling technique, Int. J. Comput. Appl. 112 (9) (2015) 19–23.
[10] S. Bhardwaj, S. Srivastava, M. Hanmandlu, J.R.P. Gupta, GFM-based methods for speaker identification, IEEE Trans. Cybern. 43 (3) (2013) 1047–1058.
[11] L. Schmidt, M. Sharifi, I.L. Moreno, Large-scale speaker identification, in: IEEE International Conference on Acoustics, Speech and Signal Processing (ICASSP), 2014, pp. 1650–1654.
[12] Y. Hu, D. Wu, A. Nucci, Fuzzy-clustering-based decision tree approach for large population speaker identification, IEEE Trans. Audio Speech Lang. Process. 21 (4) (2013) 762–774.
[13] C. Kim, R.M. Stern, Power-normalized cepstral coefficients PNCC for robust speech recognition, in: IEEE International Conference on Acoustics, Speech and Signal Processing (ICASSP), 2012, pp. 4101–4104.

[14] M.T.S. Al-Kaltakchi, W.L. Woo, S.S. Dlay, J.A. Chambers, Study of statistical robust closed set speaker identification with feature and score-based fusion, in: IEEE Statistical Signal Processing Workshop (SSP), 2016, pp. 1–5.

[15] M.T.S. Al-Kaltakchi, W.L. Woo, S.S. Dlay, J.A. Chambers, Study of fusion strategies and exploiting the combination of MFCC and PNCC features for robust biometric speaker identification, in: 4th IEEE International Conference on Biometrics and Forensics (IWBF), 2016, pp. 1–6.

[16] A. Maesa, F. Garzia, M. Scarpiniti, R. Cusani, Text independent automatic speaker recognition system using MEL-frequency cepstrum coefficient and Gaussian mixture models, J. Inf. Secur. 3 (04) (2012) 335–340.

[17] G. Nijhawan, M.K. Soni, A new design approach for speaker recognition using MFCC and VAD, Int. J. Image Graph. Signal Process. 5 (9) (2013) 43–49.

[18] E. Ambikairajah, J.M.K. Kua, V. Sethu, H. Li, PNCC-Ivector-SRC based speaker verification, in: Asia-Pacific Signal & Information Processing Association Annual Summit and Conference (APSIPA ASC), 2012, pp. 1–7.

[19] M.G. Sumithra, A.K. Devika, A study on feature extraction techniques for text independent speaker identification, in: IEEE International Conference on Computer Communication and Informatics (ICCCI), 2012, pp. 1–5.

[20] I. Trabelsi, D.B. Ayed, On the use of different feature extraction methods for linear and non linear kernels, in: 6th IEEE International Conference on Sciences of Electronics, Technologies of Information and Telecommunications (SETIT), 2012, pp. 797–802.

[21] C. Kim, R.M. Stern, Feature extraction for robust speech recognition based on maximizing the sharpness of the power distribution and on power flooring, in: IEEE International Conference on Acoustics Speech and Signal Processing (ICASSP), 2010, pp. 4574–4577.

[22] K. Kumar, C. Kim, R.M. Stern, Delta-spectral cepstral coefficients for robust speech recognition, in: IEEE International Conference on Acoustics, Speech and Signal Processing (ICASSP), 2011, pp. 4784–4787.

[23] J. Pelecanos, S. Sridharan, Feature warping for robust speaker verification, in: The Speaker Recognition Workshop Crete, Greece, 2001.

[24] H. Beigi, Fundamentals of Speaker Recognition, Springer Science & Business Media, 2011.

[25] N.V. Prasad, S. Umesh, Improved cepstral mean and variance normalization using Bayesian framework, in: IEEE Workshop on Automatic Speech Recognition and Understanding (ASRU), 2013, pp. 156–161.

[26] D.A. Reynolds, T.F. Quatieri, R.B. Dunn, Speaker verification using adapted Gaussian mixture models, Digit. Signal Process. 10 (1) (2000) 19–41.

[27] D. Reynolds, R.C. Rose, Robust text-independent speaker identification using Gaussian mixture speaker models, IEEE Trans. Speech Audio Process. 3 (1) (1995) 72–83.

[28] D. Reynolds, Large population speaker identification using clean and telephone speech, IEEE Signal Process. Lett. 2 (3) (1995) 46–48.

Further reading

[29] V.R. Apsingekar, P.L. De Leon, Support vector machine based speaker identification systems using GMM parameters, in: IEEE Conference Record of the Forty-Third Asilomar Conference on Signals, Systems and Computers, 2009, pp. 1766–1769.

[30] S.S. Nidhyananthan, R. Kumari, G. Jaffino, Robust speaker identification using vocal source information, in: IEEE International Conference on Devices, Circuits and Systems (ICDCS), 2012, pp. 182–186.

CHAPTER 8

Analysis of machine learning algorithms for audio event classification using Mel-frequency cepstral coefficients

J. Sangeetha[a], R. Hariprasad[b], and S. Subhiksha[c]
[a]Srinivasa Ramanujan Centre, SASTRA Deemed University, Kumbakonam, India
[b]Zoho Corporation, Chennai, India
[c]Department of IT, School of Computing, SASTRA Deemed University, Tirumalaisamudhram, India

8.1 Introduction

Data classification is a classic problem that has spurred numerous solutions in a variety of fields. Image classification has revolutionized the crime industry [1]. Traffic congestion is decreasing due to a well-established network of image recognizing nodes [2]. The medical field has progressed due to the vast amount of data available and the power to analyze the same. For example, cancer cells are easily identified, tumors are located faster, and heart blocks are discovered earlier [3]. Technology is making life easier. However, as always, there are lingering issues that need to be addressed. One of which is the security of women in public places. For example, there are "blind spots" to closed-circuit television (CCTV) cameras, and strict privacy laws prevent cameras from being placed in certain spaces.

Audio classification is an interdisciplinary branch of data analytics involving signal processing and machine learning methods. This specific classification problem is faced in places with high interference, loud environments, sound mixers, loud speakers, parliaments, community gatherings, music festivals, long-distance signal transmitters, space exploration satellites, and so on. This classification problem has been tackled in many places and fruitful results have been obtained [4, 5]. Audio classification, for a normal user, comes in handy to analyze multimedia data. Over the years, multimedia data has increased exponentially. Having solid audio analysis tools will help make life easier.

Applied Speech Processing
https://doi.org/10.1016/B978-0-12-823898-1.00009-6

© 2021 Elsevier Inc.
All rights reserved.

175

Audio signals are often received as streams of data. A data stream is a collection/list of continuous data sent over a period of time. The amount of data received through a stream is often much larger than the amount of time available to process it. Data fetched through streams are volatile and if tweaked properly can fetch amazing analytical results. Traditional algorithms to process streams include Bloom filters, Flajolet-Martin algorithm, and others. Stream processing is also gaining importance because of the sensitivity of time associated with data processing. Streams of data, often not less than a million bytes per minute, contain very valuable information, which when synthesized properly give us precious knowledge to tackle many real-time problems.

Foremost among these problems, as mentioned earlier, is the security of women in public spaces. There are places in countries like India where it is difficult for women to safely travel. These places often do not contain proper surveillance (e.g., cameras). This work focuses on a model intended to record and classify audio events and arrive at quick solutions to increase the security of women across the world. This work will be extended to non-destructive and intrusive monitoring of various events such as heart attacks and forest fires.

The rest of the chapter is organized as follows. Section 8.2 presents the literature survey. Section 8.3 discusses feature extraction. Section 8.4 examines machine learning techniques, and Section 8.5 presents experimental results and concluding statements.

8.2 Literature survey

Sound event detection (SED) is characterized as the undertaking of dissecting a constant sound sign to remove a depiction of sound occasions happening in the stream of sound. This depiction is regularly communicated as a name that denotes the beginning, closure, and nature (e.g., youngsters crying, spoons, wind chimes jingling) of the happening sound. Specifically, in multiname SED it is expected that more than one occasion can be dynamic (and ought to be distinguished) at once, in this manner predicting the covering of at least two of these marks this issue as a "blend issue" and it is typically not simple to understand basically, because of the superimposition (in the sound ghastly space) of energies having a place with the various occasions and to the nearness of acoustic nonidealities, for example, commotion and resonation [6, 7].

As we can see from Refs. [8, 9], hidden Markov models (HMMs) have been generally utilized for the purpose of demonstrating acoustic occasions in an SED framework. Lately, new ways to deal with SED have been proposed, denoting a particular pattern toward the utilization of counterfeit neural systems. A fascinating correlation between computational expenses of various frameworks is computed in Ref. [10] suggesting that Artificial Neural Networks (ANN) can accomplish top execution based on precision at the expense of being the most computationally costly methodology. A good example of such execution is given in Ref. [5], where distinctive ANNs were prepared on a major video dataset and afterward utilized for various degrees, among which we can likewise discover SED. For a more extensive diagram of the latest SED procedures, refer to the complete examination done by Sharan et al. in Ref. [11].

Audio event classification (AEC) also requires a viable solution to the problem of automatic language identification. A language identification application has the capacity to automatically adjust a discourse-based instrument, for example, Internet banking or data recovery, to the local language of the client. With the advent of the Internet, we have become a global society in which there are many individuals communicating and working in a wide variety of languages and dialects. Multilingual situations arise in many areas of life, including politics, the military, business, and travel [12]. Applications that recognize language are often used to book lodging, orchestrate gatherings, or make travel plans. They are also useful for call centers. For example, the language recognition application can be programmed to transfer calls to fluent speakers of the language of the caller, if necessary [13]. A programmed language recognition application could likewise fill in as a front-end for a multilanguage interpretation framework in which the information discourse is in one of a few dialects. In multilingual conditions in content autonomous language distinguishing applications, phonemes and other sub-word units alone are not adequate prompts; the application should check the sentence in general to decide the "acoustic mark" of the language, the attribute that makes one language sound unmistakable from another. To interpret the acoustic mark of a language requires data-specific acoustic phonetics, prosodic, phonotactics, jargon, and so on. A language recognizable application framework normally attempts to extricate good-quality phonetic data from the expressed articulation and use it to separate among shut set dialects [14–18]. Some more proposed methodologies are listed below.

S. no.	Title of paper	Classifiers used	Pros/cons	Publication year
1	Sound Event Detection in Real Life Audio based Neural Network Approach [19]	ANN and a binaural algorithm	+ Voice activity detection + Binaural features enhance F1 and ER scores + Short-time Fourier transform to split frames	2017
2	Learning Deep Audio Features for Video Analysis: AENet [20]	Deep convolutional neural network	+ Large temporal input field + Vocal Tract Length Perturbation (VTLP) for data augmentation + Equalized Mixture Data Augmentation (EMDA) + Effectiveness of VTLP and EMDA is significant along with large temporal input datasets	2017
3	Hidden Markov Models and Audio-Concept Features for Multimedia Event Detection [21]	Hidden Markov models, hierarchical deep neural network	+ Temporal correlation of environments considered + Gaussian mixture model – Does not explain detailed audio streams	2014
4	MFCC Feature with Different Machine Learning Techniques for Acoustic Scene Classification [22]	KNN ID3 SVM Linear discriminant analysis	+ Detailed explanations of MFCC + Feature extraction process detailed	2018
5	Sequence to Sequence Auto—Encoders For Unsupervised Learning [23]	Recurrent sequence-to-sequence auto encoder	+ Largely unsupervised learning algorithm + Recurrent sequence encoders	2017

8.3 Feature extraction for audio classification

The Mel-frequency cepstral coefficient (MFCC) is a component extraction method often used for problems involving speaker acknowledgment systems. Bridle and Brown recommended the element extraction technique in 1974. Mermelstein subsequently proposed it in 1976. Audio classification problems can be analyzed by MFCC because it has the potential to imitate the tumult and pitch of human sound-related structure in a logarithmic vision. Through preserving the music and vital reappearance of sound, it clears the speaker subordinate characters. To recognize active conversation, the MFCC incorporates the disparity in the component vectors. The systematic procedure for producing the MFCC is illustrated below.

8.3.1 Preemphasis

To enhance the strength of high-frequency signals, the preemphasis filter has been proposed. At the time of the Fourier transform process the preemphasis filter is utilized to vary the reappearance range to eliminate numerical problems. To obtain high-frequency standards, these ranges are adjusted. This shows that high-frequency components have lesser magnitudes than low-frequency components.

8.3.2 Framing

Here the signals are segregated into a small number of frames. After a short period, the signal frequency changes. If the entire signal is involved in the Fourier transform process, then the signal's frequency contours with respect to time are ignored. We can maintain a stable frequency in the signal for a long time in order to eliminate this problem. By connecting the adjoining outlines, we can determine a calculation of the frequency contours of the signal through the Fourier change over this little time slot. A frame size of 50 ms and a frame step of 50 ms are taken for the calculation.

8.3.3 Windowing

Windowing is applied to reduce spectral distortion and eliminate the discontinuities of each frame at both sides in the signals. The Hamming window function is used to maintain the balance between frequency resolution and dynamic range of the signal.

8.3.4 Discrete Fourier transforms (DFTs)

The following equation is used to determine the DFT of a frame:

$$S_i(k) = \sum_{n=1}^{N} s_i(n)h(n)e^{-j2\pi kn/N} \quad 1 \le k \le K$$

Here $h(n)$ is a Hamming window of N size, and K is the DFT length. The periodogram-based power spectral estimate for the speech frame $s_i(n)$ is the periodogram that estimates the power spectrum:

$$P_i(k) = \frac{1}{N}|S_i(k)|^2$$

Complex Fourier transform's absolute value is obtained and then squared. The first 257 coefficients are taken by performing 512-point DFT.

8.3.5 Mel-scaled filter banks

To segregate the recurrence groups, the filter banks are visualized by relating triangular channels to the Mel scale. We can change between Mel (m) and Hertz (f) by means of the following equations.

The method for change to Mel scale from frequency is:

$$M(f) = 1125 \ \ln(1 + f/700) \tag{8.1}$$

Use the following to move from Mel back to frequency:

$$M^{-1}(m) = 700(\exp(m/1125) - 1) \tag{8.2}$$

Each filer in the bank is triangular, having response as one at the center and linearly decreasing to zero until it moves center frequencies of the adjacent two filters as shown in Fig. 8.1. MFCC is calculated from a sample window of 512 points and comprises 26 Cepstral coefficients, 13 first-order and 13 second-order derivatives. Only the first 13 coefficients are considered because they represents the spectra's envelope and reduce feature-space dimensionality.

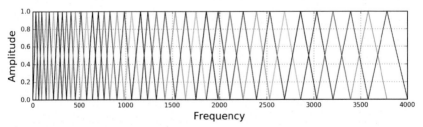

Fig. 8.1 Mel scale in MFCC extraction [24].

8.4 Machine learning techniques

8.4.1 Support vector machines (SVMs)

In the field of speech and audio recognition, many supervised learning algorithms have proved to be accurate. Support vector machines (SVMs) are promising classification algorithms with a wide range of customizable parameters to obtain necessary results. The basic principle of SVMs is choosing samples from various classes (support vectors) and structuring a linear function to split them accordingly. SVMs are classifiers that perform classification tasks through utilizing hyperplanes in multidimensional space. They use parameters to identify decision planes and data planes and compute distance between them to obtain the most probable class fitting in a hyperplane (Fig. 8.2).

The Python kernel trick can be employed in traditional SVM to obtain different shapes of hyperplanes as against the traditional line as a hyperplane. By performing the kernel trick, we obtain a model that is a better fit for the dataset. Some of the commonly used kernel tricks are radial basis function (RBF) kernel, linear kernel, and polynomial kernel (poly kernel).

Depending on the relationship between the features, any one of these kernels can give a better output. Based on which kernel gives the better output, we can identify the relationship between the features and the output variable [25]. In SVM, a line is drawn to separate the different class labels. If class label "a" exists on one side of the line, class label "b" exists on the

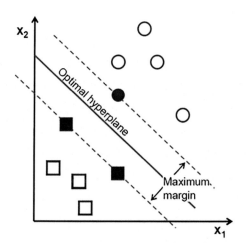

Fig. 8.2 SVM hyperplane.

other. Accuracy is calculated based on how well the SVM can segregate the class labels to either side of the line.

The basic principle of an SVM with a poly kernel remains the same as an SVM with a linear kernel or an SVM with an RBF kernel. Here the kernel trick is employed to make the hyperplane fit the dataset better. In a polynomial kernel, the line that separates the different class labels is in the higher dimension. This kernel fits the dataset best when it is used on nonlinear data. The value obtained by this classifier depends on the Euclidean distance between two points in space or the Euclidean distance between the origin and the point under consideration.

Apart from the kernel trick, there are some parameters in the SVM that can be tuned to obtain better results. These parameters are called "gamma" and "C" parameters. The C parameter represents regularization. A greater value of C implies a decrease in the margins of the hyperplane. Here the amounts of examples that are classified wrongly is reduced. Greater regularization means that the model that is obtained is a better fit for the dataset. Gamma value determines the points to take into consideration during the building of the separating line. A small value of gamma means the points farthest away from the separating line are taken into consideration. On the contrary, bigger gamma values mean that only the points closest to the line are taken into consideration.

8.4.2 Logistic regression

Logistic regression uses a logit function as its base function and all values are bound between 0 and 1. It is very useful for classification when compared to linear regression. Though logistic regression has "regression" in its name, it is not a regression method. On the contrary, it is a classification method. There are three distinct types of logistic regression as follows:

Ordinal logistic regression: Where there are multiple class labels and each class label is in a hierarchical order (1-star rating, 2-star rating).

Binary logistic regression: Where there are only two class labels (Yes–No, True–False).

Multinomial logistic regression: Where there are multiple class labels and each of these class labels doesn't have a logical order (Pass–Fail–Absent–Repeat) [26].

8.4.3 Decision tree

This classifier builds a tree to classify the various class labels present in the dataset. The various features present in the dataset are considered and the probabilities associated with those algorithms are found. Gini index is then calculated based on the probabilities obtained. Whichever feature has the greatest Gini index is selected as the root node. The corresponding nodes are chosen based on the decreasing order of the Gini index. Once a node has been chosen, the variables are divided into the various class labels present in the data. In Python, for splitting the node, entropy can also be used as a criterion [27].

8.5 Experimental results and discussion

8.5.1 Dataset

We used the Research on Detection and Classification of Acoustic Scenes and Events (DCASE) challenge dataset. The TAU Urban Acoustic Scenes (2019) dataset consists of soundtracks from a variety of acoustic scenes. This corpus extends the TUT Urban Acoustic Scenes (2018) corpus with six other cities. As such, it consists of 12 big European cities. For every scene class, soundtracks were prepared in diverse locations; all the audio footage files are 5–6 min in length. The original footages were divided into 10-s segments given in separate files. Information about the footages includes city, acoustic scene class, and recording location.

8.5.2 Experimental setup

We obtained audio files and used Python code to extract the characteristics of the audio files. We then scaled the obtained data using the MIN MAX scaler in Python. We loaded the class labels as a separate file. The dataset contained 193 features and 9 class labels, which we split into 2 parts. One part contained 80% of the original dataset and was used as the training dataset, and the other 20% was used as a test dataset.

Implementation of the algorithm was achieved through the following classifiers: SVM with linear, polynomial, and RBF kernels; decision tree; and logistic regression. Every algorithm uses parameters such as soft margin parameter C for SVM, Gini index for the decision tree classifier and multiclass parameter.

8.5.2.1 Support vector machine with linear kernel

This algorithm uses a line to separate the data belonging to different class labels. When the linear kernel is used, the dot product of the input and the support vector are calculated [28]. The regularization parameter was also adjusted to acquire greater accuracy. The C value was set to 0.9.

The probability value was set to true so that a receiver operating characteristic (ROC) curve could be drawn. Results showed accuracy of 82.5%, calculated by drawing the confusion matrix (Fig. 8.3).

8.5.2.2 Support vector machine with polynomial kernel

The basic principle of SVM with a poly kernel is the same as SVM with a linear kernel or SVM with an RBF kernel. Here the kernel trick is employed to make the hyperplane fit the dataset better. In the polynomial kernel, the line that separates the different class labels is in the higher dimension. Before MIN MAX scaling was done on the data, the polynomial kernel gave greater accuracy than the linear kernel did. After scaling when the polynomial kernel was used, results showed an accuracy of 26.25%. This suggests that this kernel is not the best fit for this dataset (Fig. 8.4).

8.5.2.3 Support vector machine with radial basis function kernel

This kernel fits the dataset best when it is used on nonlinear data. The value obtained by this classifier depends on the Euclidean distance between two points in space or the Euclidean distance between the origin and the point under consideration. Results showed an accuracy of 43.75%. This model performs better than the SVM poly model, but the accuracy is considerably less when compared to that given by SVM with a linear kernel.

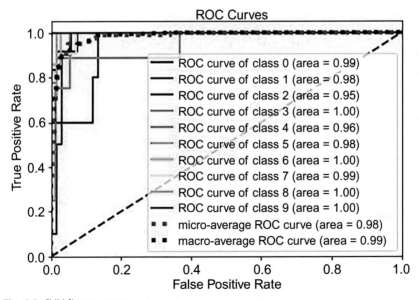

Fig. 8.3 SVM linear accuracy.

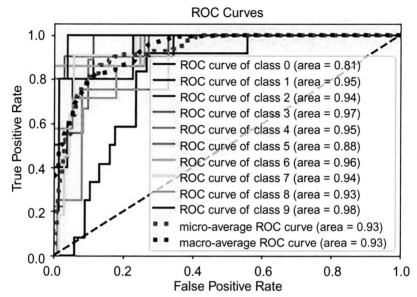

Fig. 8.4 SVM poly accuracy.

8.5.2.4 Decision tree algorithm

A decision tree was built using Gini index as a criterion. Here the Gini index is used to decide the feature based on which the decision tree is to be split. The algorithm was designed in such a way that there is a minimum of 10 samples per leaf. Decrease in impurity has a threshold value and this value decides if a node has to be split or not. In this case, the value was set to 0.01. This algorithm resulted in an accuracy of 56.25%. This was greater than the accuracy given by the SVM poly kernel, but was much less than the accuracy obtained by the SVM linear kernel (Figs. 8.5 and 8.6).

8.5.2.5 Logistic regression

The multiclass parameter available in logistic regression in Python was set to multinomial logistic regression. Using a multinomial mode requires a mechanism to handle multinomial loss. Therefore, we used the Newton–CG solver. Results showed accuracy of 81.25%. This accuracy was very close to the accuracy given by the SVM linear classifier (Fig. 8.7).

The summary of the names of all the models along with their accuracy is given below.

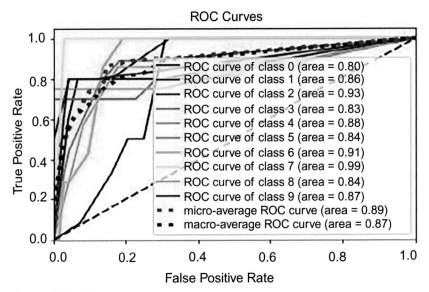

Fig. 8.5 SVM RBF accuracy.

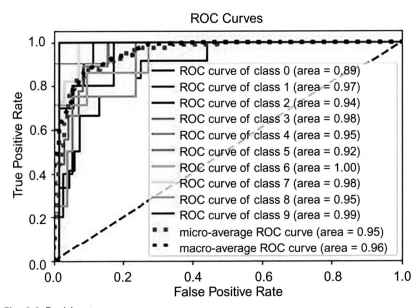

Fig. 8.6 Decision tree accuracy.

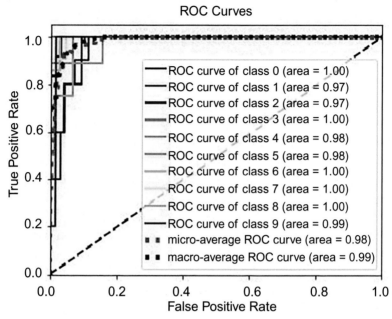

Fig. 8.7 Logistic regression accuracy.

Model name	Accuracy (in percentage)
SVM linear	82.5%
SVM poly	26.25%
SVM Rbf	43.75%
Decision tree	56.25%
Logistic regression	81.25%

8.6 Conclusion

In the future, we intend to implement an ensemble model involving logistic regression as the base classifier. Following this, we plan to train a convolutional recurrent neural network with deep layers. We predict that the new model will have greater accuracy with positive effects from the logistic regression model and deep neural networks. Having deep neural networks opens up wide avenues of feature analysis. In addition, we plan to scrutinize the scalar we used to preserve the prosodic integrity of the audio data. Feature extraction will be modified along with MFCC features to obtain a custom extraction algorithm that may prove to have greater accuracy for AEC.

References

[1] D.K. Tayal, A. Jain, S. Arora, S. Agarwal, T. Gupta, N. Tyagi, Crime detection and criminal identification in India using data mining techniques, AI Soc. 30 (1) (2015) 117–127.

[2] C.L. Wan, K.W. Dickinson, Road traffic monitoring using image processing—a survey of systems, techniques and applications, IFAC Proc. Vol. 23 (2) (1990) 27–34.

[3] P. Arena, A. Basile, M. Bucolo, L. Fortuna, Image processing for medical diagnosis using CNN, Nucl. Instrum. Methods Phys. Res., Sect. A 497 (1) (2003) 174–178.

[4] F. Rong, Audio classification method based on machine learning, in: 2016 International Conference on Intelligent Transportation, Big Data & Smart City (ICITBS), December, IEEE, 2016, pp. 81–84.

[5] S. Hershey, S. Chaudhuri, D.P. Ellis, J.F. Gemmeke, A. Jansen, R.C. Moore, M. Plakal, D. Platt, R.A. Saurous, B. Seybold, M. Slaney, CNN architectures for large-scale audio classification, in: 2017 IEEE International Conference on Acoustics, Speech and Signal Processing (ICASSP), March, IEEE, 2017, pp. 131–135.

[6] A. Mesaros, T. Heittola, A. Eronen, T. Virtanen, Acoustic event detection in real life recordings, in: 2010 18th European Signal Processing Conference, August, IEEE, 2010, pp. 1267–1271.

[7] R. Rotili, E. Principi, S. Squartini, B. Schuller, A real-time speech enhancement framework in noisy and reverberated acoustic scenarios, Cogn. Comput. 5 (4) (2013) 504–516.

[8] T. Heittola, A. Mesaros, A. Eronen, T. Virtanen, Audio context recognition using audio event histograms, in: 2010 18th European Signal Processing Conference, August, IEEE, 2010, pp. 1272–1276.

[9] Y.T. Peng, C.Y. Lin, M.T. Sun, K.C. Tsai, Healthcare audio event classification using hidden Markov models and hierarchical hidden Markov models, in: 2009 IEEE International Conference on Multimedia and Expo, June, IEEE, 2009, pp. 1218–1221.

[10] S. Sigtia, A.M. Stark, S. Krstulović, M.D. Plumbley, Automatic environmental sound recognition: performance versus computational cost, IEEE/ACM Trans. Audio Speech Lang. Process. 24 (11) (2016) 2096–2107.

[11] R.V. Sharan, T.J. Moir, An overview of applications and advancements in automatic sound recognition, Neurocomputing 200 (2016) 22–34.

[12] M. Adda-Decker, L. Lamel, The use of lexica in automatic speech recognition, in: Lexicon Development for Speech and Language Processing, Springer, Dordrecht, 2000, pp. 235–266.

[13] Y.K. Muthusamy, E. Barnard, R.A. Cole, Automatic language identification: a review/tutorial, IEEE Signal Process. Mag. 11 (4) (1994) 33–41.

[14] E. Noor, H. Aronowitz, Efficient language identification using anchor models and support vector machines, in: 2006 IEEE Odyssey—The Speaker and Language Recognition Workshop, June, IEEE, 2006, pp. 1–6.

[15] N. Yang, et al., Emotional state recognition for AI smart home assistants using Mel-frequency cepstral coefficient features, J. Intell. Fuzzy Syst. (2019).

[16] S. Sen, A. Dutta, N. Dey, Audio Processing and Speech Recognition: Concepts, Techniques and Research Overviews, Springer, 2019.

[17] N. Dey (Ed.), Intelligent Speech Signal Processing, Academic Press, 2019.

[18] S. Sen, A. Dutta, N. Dey, Speech processing and recognition system, in: Audio Processing and Speech Recognition, Springer, Singapore, 2019, pp. 13–43.

[19] N. Dey, A.S. Ashour, W.S. Mohamed, N.G. Nguyen, Acoustic sensors in biomedical applications, in: Acoustic Sensors for Biomedical Applications, Springer, Cham, 2019, pp. 43–47.

[20] M. Valenti, D. Tonelli, F. Vesperini, E. Principi, S. Squartini, A neural network approach for sound event detection in real life audio, in: 2017 25th European Signal Processing Conference (EUSIPCO), IEEE, 2017, pp. 2754–2758.

[21] N. Takahashi, M. Gygli, L. Van Gool, Aenet: learning deep audio features for video analysis, IEEE Trans. Multimedia 20 (3) (2017) 513–524.

[22] B. Elizalde, M. Ravanelli, K. Ni, D. Borth, G. Friedland, Audio-concept features and hidden Markov models for multimedia event detection, in: SLAM@ INTERSPEECH, September, 2014, pp. 3–8.

[23] M.M. Oo, Comparative study of MFCC feature with different machine learning techniques in acoustic scene classification, Int. J. Res. Eng. (2018) 2348–7860.

[24] J. Sangeetha, S. Jothilakshmi, A novel spoken keyword spotting system using support vector machine, in: Engineering Applications of Artificial Intelligence, vol. 36, Elsevier publication, 2014, pp. 287–293. SCI—E, SCOPUS, indexed—IF: 2.986.

[25] S. Amiriparian, M. Freitag, N. Cummins, B. Schuller, Sequence to sequence autoencoders for unsupervised representation learning from audio, in: Proc. of the DCASE 2017 Workshop, November, 2017.

[26] https://medium.com/machine-learning-101/chapter-2-svm-support-vector-machine-theory-f0812effc72#:~:targetText=0.,hyperplane%20which%20categorizes%20new%20examples (Chapter 2: SVM (Support Vector Machine)—Theory, 2020).

[27] https://medium.com/greyatom/decision-trees-a-simple-way-to-visualize-a-decision-dc506a403aeb.

[28] https://towardsdatascience.com/logistic-regression-detailed-overview-46c4da4303bc.

Index

Note: Page numbers followed by *f* indicate figures and *t* indicate tables.

Printed in the United States
By Bookmasters